Regulation of Air Transport

Ruwantissa Abeyratne

Regulation of Air Transport

The Slumbering Sentinels

Ruwantissa Abeyratne
Montreal
Québec
Canada

ISBN 978-3-319-01040-3   ISBN 978-3-319-01041-0 (eBook)
DOI 10.1007/978-3-319-01041-0
Springer Cham Heidelberg New York Dordrecht London

Library of Congress Control Number: 2013948500

Printed on acid-free paper

Springer is part of Springer Science+Business Media (www.springer.com)

Preface

The airline and airport industry combined employs 4.3 million people. Additionally, engine and airframe component manufacturers employ 73,000. These industries indirectly provide employment for 5.8 million people, mostly within the supply chain that services these industries. In 2010, 2.6 million flights were operated every month providing 317 million seats. Aircraft departures are forecast to grow from 30 million in early 2013 to 60 million by 2030. Figures of the International Air Transport Association (IATA) for 2010 indicate that there were 1,568 commercial airlines; 23,844 aircraft in commercial service (19,203 jets and 4,641 turboprops); 3,846 airports servicing these aircraft which operated 26.7 million commercial aircraft movements. Airports Council International (ACI) has reported that at these airports, in 2011, worldwide traffic reached 5.4 billion passengers. Furthermore, during the first half of 2012, traffic managed to grow by a respectable 5 % in spite of global uncertainty. Regionally, strong passenger traffic growth continues in the Middle East, Latin America, Africa, and Asia, while the mature markets of North America and Western Europe growth are somewhat stalled. Airports worldwide are forecast to handle over 10 billion passengers by 2030.

At the time this book was being written, *Airbus Industrie* had reported that it had delivered 588 aircraft in 2012. It was going ahead with the A 350—Airbus's answer to the Boeing 787 Dreamliner—which was being questioned for defects. *Airbus Industrie*, it was reported, had made a profit of 1.5 billion US dollars in 2012. Over the past decade, airlines carried 70 % more people than were carried in the previous decade, and this figure is expected to double by 2030. At its 12th Air Navigation Conference held in November 2012, the International Civil Aviation Organization (ICAO) released its new Global Air Navigation Plan, which would cope with the congestion in the skies brought about by the exponential traffic growth. Yet, not much progress had been made in the past 10 years to provide a sense of purpose and direction to the economic regulation of air transport.

Although, like many other businesses, the air transport business is sometimes beset by surprises such as unexpected economic crashes, wide-scale terrorist activity, pandemics, and natural disasters, those in the business know or ought to know the nature of the business and have the know-how to adapt to the vicissitudes of cyclical business patterns that apply. However, they are precluded from doing so by the very nature of the air transport industry, which is driven by government policy, competition, and a race to the finish line ahead of others. Airlines are attracted by the claims of manufacturers who sport new aircraft that are more efficient, emit less greenhouse gases, and cause less noise than their competitors' aircraft. They order aircraft that, when ultimately delivered, are unable to cope with economic factors that have overtaken the circumstances prevailing when they were ordered.

Once every 10 years ICAO holds a worldwide air transport conference. The most recent of such events—the 6th Worldwide Air Transport Conference (ATConf/6)—was held in Montreal from 18 to 22 March 2013. At this conference, delegates mulled over "achievements" since the 5th Worldwide Air Transport Conference (ATConf/5), held 10 years earlier in Montreal from 24 to 29 March 2003. They noted that ATConf/5 had adopted a Declaration on Global Principles for the Liberalization of International Air Transport, 14 Conclusions and 2 Recommendations, i.e., one on the liberalization of air carrier ownership and control and the other on the future role of ICAO, including its relations with the World Trade Organization (WTO-OMC), and seven model clauses for use at States' discretion in air services agreements.

In addition, ATConf/5 adopted 67 Conclusions for inclusion in its report, which covered the full spectrum of topics and issues on the agenda. Among the Conclusions was the conference's approval for the concept and content of the Template Air Services Agreements (TASAs), one for bilateral and the other for regional situations, for use in air transport relationships. Throughout the Declaration, Recommendations, and Conclusions are numerous references to actions to be taken by States and by ICAO.

Based on these (the Declaration, Conclusions, and Recommendations), ATConf/6 decided, as further action to "meet the needs of the people of the world for safe, regular, economical and efficient air transport" (as prescribed in the Convention on International Civil Aviation), to continue to assist States in their liberalization efforts by enhancing the "market place" facility offered to States; continue to update the TASAs to keep pace with regulatory evolution; undertake and promote the development of additional training courses, regional seminars, or similar activities for the benefit of States, in accordance with available resources; continue to monitor regulatory developments, conduct studies on major issues of global importance, and provide policy guidance and assistance to States; and continue to develop relevant databases such as the Database of the World's Air Services Agreements, as well as case studies of liberalization experiences.

The questions posed by this book are as follows: Are these "clerical and administrative tasks" for ICAO that were decided on by ATConf/6 (and other such preceding conferences) sufficient to meet the needs of the people of the world for safe, regular, economical, and efficient air transport? Should ICAO not think out of its 67-year-old box and be a beacon to air transport regulators? In other words, shouldn't the bottom line of ICAO's meaning and purpose in the field of air transport be to analyze trends and guide the air transport industry instead of continuing to act as merely a forum for global practitioners to gather and update information on their respective countries' policies for air transport? Shouldn't ICAO provide direction as do other agencies of the United Nations?

These questions are easily answered by a sensible response to the question "what are the needs of the people in this regard?" For instance, the World Bank assembles record funding to reduce poverty. The World Health Organization (WHO) is the directing and coordinating authority for health within the United Nations system. It is responsible for providing leadership on global health matters, shaping the health research agenda, setting norms and standards, articulating evidence-based policy options, providing technical support to countries, and monitoring and assessing health trends. The International Labour Organization (ILO) is the international organization responsible for drawing up and overseeing international labour standards. The International Maritime Organization (IMO) is the United Nations specialized agency with responsibility for the safety and security of shipping and the prevention of marine pollution by ships.

The word "responsibility" does not occur anywhere in ICAO's mission in air transport, as it does in other specialized agencies of the United Nations. In its web page, ICAO identifies itself as a specialized agency of the United Nations that *promotes* the safe and orderly development of international civil aviation through-out the world. It sets standards and regulations necessary for aviation safety, security, efficiency, and regularity, as well as for aviation environmental protection. Its mission is to be the global forum for civil aviation—in other words, a meeting place. Its only task is to achieve safe, secure, and sustainable development of civil aviation through the cooperation of its Member States. Again, in other words, ICAO is but an interlocutor (someone who takes part in a conversation) in global air transport, and if States do not cooperate in a particular issue, ICAO is destitute of the ability to show direction and persuade its members to follow it.

This book addresses ICAO's inability, unlike most other specialized agencies in their missions, to make a tangible difference to air transport development, through a discussion of key issues affecting the air transport industry. It will also inquire into the future of air transport regulation.

Montreal, QC, Canada   Ruwantissa Abeyratne
July 2013

Contents

1   **The Nature and Role of ICAO** 1
    1.1   The Problem 1
    1.2   The Solution 6
    1.3   Competition 11
    1.4   The ICAO-IATA Synergy 14
    1.5   Governance 19
    References 25

2   **Competition in Air Transport** 27
    2.1   The Safeguards Anomaly 29
    2.2   Competition in Air Transport 36
    2.3   Ownership and Control of Airlines 39
    2.4   Existing Guidelines 47
    2.5   A Compromise 52
    References 56

3   **Connectivity** 57
    3.1   The Declaration 57
    3.2   The Tourism Connection 59
    3.3   Airport Slots 65
        3.3.1   Night Curfews 73
    References 76

4   **The Facilitation Connection** 77
    4.1   Carriage of Persons 79
    4.2   Carriage of Cargo 82
    References 83

5   **Consumer Rights** 85
    5.1   Consumer Protection 85
    5.2   Consumer Rights 92
    References 101

6   Airlines: The Other Side of the Coin   103
    6.1   Airlines and Connectivity   103
    6.2   Charges and Taxes: A Burden on Airlines   107
           6.2.1   Definition of Charge   109
           6.2.2   Marginal Cost Pricing   113
           6.2.3   The Single Till   115
           6.2.4   CEANS 2008   121
           6.2.5   ICAO's Policies on Airport Charges   129
    6.3   Economic Regulation and Security of Airlines: The Importance
           of Cargo   136
           6.3.1   Economic Aspects of Security   138
           6.3.2   Cargo Security   139
           6.3.3   The Risk Based Approach   143
    6.4   Economic Regulation of Air Navigation Services Providers   150
    References   154

7   The Black Swan Effect   157
    7.1   Epiphenomena   159
    7.2   The Low Fare Business Model   161
    7.3   A Sensible Approach to Fragility?   167
    References   172

8   Outcome of the Sixth Air Transport Conference (ATConf/6)   173
    8.1   Discussions and Achievements of ATConf/6   174
           8.1.1   Market Access   175
           8.1.2   Ownership and Control   180
           8.1.3   Consumer Protection   184
           8.1.4   Fair Competition   186
           8.1.5   Safeguards in the Liberalization Process   189
           8.1.6   Taxation   193
           8.1.7   Economics of Airports and Air Navigation Services   194
    8.2   Implementation of ICAO's Policies and Guidance   197
    8.3   A New Annex?   199
    References   202

9   Conclusion   203
    References   210

Index   211

Chapter 1
The Nature and Role of ICAO

1.1 The Problem

ICAO[1] has its genesis in Article 57 of the Charter of the United Nations which states:

> The various specialized agencies, established by intergovernmental agreement and having wide international responsibilities, as defined in their basic instruments, in economic, social, cultural, educational, health, and related fields, shall be brought into relationship with the United Nations in accordance with the provisions of Article 63.
>
> Such agencies thus brought into relationship with the United Nations are hereinafter referred to as specialized agencies.[2]

Clearly, this provision gives ICAO "wide international responsibilities, as defined in their basic instruments..." In all fairness to ICAO, the basic instrument in this context—the Convention on International Civil Aviation (hereafter referred to as the Chicago Convention)—defines the aims and objectives (not mandate or responsibility) of the Organization:

> The aims and objectives of the Organization are to develop the principles and techniques of international air navigation and to foster the planning and development of international air transport.[3]

---

[1] ICAO is the specialized agency of the United Nations handling issues of international civil aviation. ICAO was established by the Convention on International Civil Aviation, signed at Chicago on 7 December 1944 (Chicago Convention). One of the overarching objectives of ICAO, as contained in Article 43 of the Convention is to foster the planning and development of international air transport so as to meet the needs of the peoples for safe, regular, efficient and economical air transport. ICAO has 191 member States, who become members of ICAO by ratifying or otherwise issuing notice of adherence to the Chicago Convention. See ICAO Doc 7300/9 Ninth Edition 2008.

[2] Charter of the United Nations and Statute of the International Court of Justice, United Nations: New York, DPI/511-40108 (3-90) 100M, Article 57.1 and 57.2.

[3] Chicago Convention, *supra*, note 1, Article 44.

R. Abeyratne, *Regulation of Air Transport*, DOI 10.1007/978-3-319-01041-0_1,

In terms of economic regulation of air transport, ICAO can only "foster the planning and development of air transport". One definition of "foster" is "to encourage or promote the development of something (typically something regarded as good)". Another is "to nurture". None of these definitions has even a hint of leadership, direction or responsibility ascribed to ICAO. At the International Civil Aviation Conference held in Chicago from November 1 to December 7, which gave rise to the Chicago Convention, some delegations suggested a Resolution which stated *inter alia*:

> We the nations and authorities represented at this international civil aviation conference, being determined that the fullest measure of cooperation should be secured in the development of air transport services between the nations of the world, believing that the unregulated development of air transport can only lead to misunderstanding and rivalries between nations. . .agree that these objectives can best be achieved by the establishment of an international air transport authority which would be responsible for the operation of air services on prescribed international trunk routes. . .

This resolution did not see the light of day.[4] As a result, after nearly 70 years we have a somewhat powerless Organization grappling with economic issues without the responsibility to lead its member States. This is indeed unfortunate, as a leading political scientist said:

> The primary aim of inter-governmental institutions is to produce some predictability about the behaviour of its members.[5]

The key to inducing such predictability is the promulgation of Standards for the conduct of member States, an ability not bestowed upon ICAO by its authorizing instrument. However, all is not lost, due to the ingenuity of the Council of ICAO. The Council, at the 8th meeting of its 196th Session in June 2012, approved ICAO's revised Vision and Mission Statements, as well as the new set of five Strategic

---

[4] Early in the discussions during the Chicago Conference of 1944 which led to the adoption of the Chicago Convention, the Committee on Multilateral Aviation Convention and International Aeronautical Body rejected the joint proposal from the New Zealand and Australian Delegations for international ownership and operation of civil air services on world trunk routes. The rejection of that proposal indicated the tendency of the Conference away from extensive international control of air services. Of the three other plans which remained before the Committee, the United States plan called for an international aviation authority with powers limited to the technical and consultative fields; the Canadian plan aimed to set up international authority with power to allocate routes, review rates, and determine frequencies of operation, but with that power curbed by specific formulae under which the authority would operate; and the United Kingdom plan proposed more discretionary power to the international authority in allocating routes, fixing rates, and determining frequencies. It was soon obvious that none of the three plans would emerge intact from the discussions and that the final Conference proposal, if agreement were reached, would be a composite of all plans. See *Proceedings of the International Civil Aviation Conference*, Chicago, Illinois, November 1 to December 7, Vol. I, The Department of State, United States Government Printing Office, Washington:1948, at 539–540.

[5] Gordenker (1991), p. 71.

Objectives of the Organization for 2014–2016.[6] Accordingly, ICAO's new vision statement is "Achieve the sustainable growth of the global civil aviation system". Its Mission Statement is "The International Civil Aviation Organization is the global forum of States for international civil aviation. ICAO develops policies, standards, undertakes compliance audits, performs studies and analyses, provides assistance and builds aviation capacity through the cooperation of Member States and stakeholders". Under these two statements come ICAO's new Strategic Objectives:

A. Safety: Enhance global civil aviation safety
B. Air Navigation Capacity and Efficiency: Increase capacity and improve efficiency of the global civil aviation system
C. Security and Facilitation: Enhance global civil aviation security and facilitation
D. Economic Development of Air Transport: Foster the development of a sound and economically-viable civil aviation system
E. Environmental Protection: Minimize the adverse environmental effects of civil aviation activities.

When one meshes the Mission Statement which states *inter alia* that ICAO develops policies, with Strategic Objective D (Economic Development of Air Transport) ICAO has legal justification to interpret the original aim in Article 44 of the Chicago Convention in a manner that justifies more responsibility and leadership in economic regulation. However, the spoiler remains the explanation of Strategic Objective D: "Foster the development of a sound and economically-viable civil aviation system". The Council has ineptly reverted back to "fostering" and that too, something called "a sound and viable civil aviation system". Apart from the ambivalence that resonates through words such as "sound" and "viable", one is bemused by the term "civil aviation system". The author has not seen any definition of this term. However, if civil aviation is considered in a systemic way, it should include all aspects of civil aviation including the aircraft involved and their manufacture and that of their components, the service providers such as ground handlers. ICAO has nothing to do with these elements.

In terms of ICAO's role in facilitating cooperation among States, the Organization is consistent with the overall role to be played by the United Nations specialized agencies as was recognized in 1945:

> The international institutions of the post war period continue to be primarily instruments for the organization of cooperation between States.[7]

However, we are in the twenty-first century and the role of a United Nations agency calls for a more proactive role. This proactive role has been epitomized in the context of responses of the United Nations specialized agency thus:

---

[6] Earlier, ICAO had four Strategic Objectives. They were: Safety—Enhance global civil aviation safety; Security—Enhance global civil aviation security; Environmental Protection—Minimize the adverse effect of global civil aviation on the environment; and Sustainable Development of Air Transport.

[7] Jenks (1945), pp. 11, 19.

The evolving role of the UN specialized agencies, those quasi-autonomous UN bodies which report to the Economic and Social Council (ECOSOC) rather than directly to the General Assembly, has received less attention, despite the efforts they have made in recent years to adapt to the new demands of emergencies of all kinds by developing new priorities and ways of working, and by playing an increasingly active part in coordination mechanisms for emergency response.[8]

In the case of ICAO the best example of a complex emergency was 9/11 when aircraft were used as weapons of mass destruction. ICAO responded with aplomb, at the 33rd Session of the Assembly (Montreal, 25 September to 5 October 2001), adopting Resolution 33/1 entitled *Declaration on misuse of civil aircraft as weapons of destruction and other terrorist acts involving civil aviation*. This Resolution, while singling out for consideration the terrorist acts which occurred in the United States on 11 September 2001, and, *inter alia*, recognizing that the new type of threat posed by terrorist organizations requires new concerted efforts and policies of cooperation on the part of States, urged all Contracting States to intensify their efforts in order to achieve the full implementation and enforcement of the multilateral conventions on aviation security, as well as of the ICAO Standards and Recommended Practices and Procedures (SARPs) relating to aviation security, to monitor such implementation, and to take within their territories appropriate additional security measures commensurate to the level of threat in order to prevent and eradicate terrorist acts involving civil aviation. The Resolution also urged all Contracting States to make contributions in the form of financial or human resources to ICAO's aviation security mechanism to support and strengthen the combat against terrorism and unlawful interference in civil aviation; called on Contracting States to agree on special funding for urgent action by ICAO in the field of aviation security; and directed the Council to develop proposals and take appropriate decisions for a more stable funding of ICAO action in the field of aviation security, including appropriate remedial action.

Resolution 33/1 directed the Council to convene, at the earliest date, if possible in the year 2001, an international high-level, ministerial conference on aviation security in Montreal with the objectives of preventing, combatting and eradicating acts of terrorism involving civil aviation; of strengthening ICAO's role in the adoption of SARPs in the field of security and the audit of their implementation; and of ensuring the necessary financial means to strengthen ICAO's AVSEC Mechanism, while providing special funding for urgent action by ICAO in the field of aviation security.

The effects of this resolution were immediately seen in certain concerted efforts made both in the United States and Europe to take immediate measures to strengthen aviation security. The European Transport and Telecommunications Council, at its meeting in Luxemburg on 16 October 2001, welcomed the proposal by the Commission for a Regulation establishing common rules in the field of civil

---

[8] White (1999), pp. 223–238.

aviation security.[9] The Council invited Member States and the European Commission to contribute to the preparation for the ICAO High Level/Ministerial Conference as referred to in Resolution 33/1. In the United States, Ms. Jane Garvey, FAA Administrator, stated on 17 October 2001 at a meeting in Washington, that the United States will start using new technology called the Computer Assisted Passenger Pre-screening System which would introduce new technologies to detect plastic weapon, and greater use of explosive detection equipment.[10] Ms. Garvey further added that Transportation Secretary Mineta had created a $20 million dollar fund to explore new technologies to improve aircraft security. These grants could be used to test any new technology that leads to safer, more secure aircraft.[11]

The operative question is, does ICAO show the same vigilance, robustness and proactivity in the economic regulation of air transport? The answer is readily available from the results of the 6th Air Transport Conference (ATConf/6) held in March 2013. At that Conference, there were several working papers discussed, but no collective strategy nor direction for future economic regulation of air transport flowed through. Even at the Pre Conference Symposium (which is now a regular practice at ICAO) held a day before the Conference the speakers spoke mostly of current trends and not about solutions or direction. Each was polarized towards its own interest (for instance UNWTO spoke of tourism and how cooperation would best work to enhance tourism and the airports did the same with their subject). This has been a trend for the past 20 years.

To the contrary, At its 12th Air Navigation Conference (AN-Conf/12) held in Montréal, from 19 to 30 November 2012, ICAO introduced its Global Air Navigation Capacity and Efficiency Plan (GANP) for the period 2013–2028. The Plan is meant to be approved by the ICAO Assembly at its sessions every 3 years. This is the fourth such Plan adopted by ICAO over the years and is based on operational objectives agreed upon by States and the aviation industry. It provides for a rolling 15 year strategic methodology and introduces Aviation System Block Upgrades (ASBUs), each of which has a 5 year time scale. The ASBUs are not overarching, nor are they an umbrella system but remain flexible modules that can be used by States in accordance with their individual operational needs. One of the salient characteristics of ASBUs is that they define technologies and procedures that are calculated to improve operational performance, particularly when the need arises for an operational problem to be solved. The ultimate aim is to achieve global harmonization and interoperability of air navigation.

Another example of ICAO proactivity was at the High Level Conference on Aviation Security, convened by ICAO from 12 to 14 September 2012 where the conference recognized that air cargo advanced information for security risk assessment is a developing area that enhances air cargo security, particularly in the context of express delivery carriers such as FEDEX, UPS, DHL Express and TNT Express who carry around 30 million shipments daily, which typically contain

---

[9] Transport and Telecommunications Council, Luxemburg, 16 October 2001.
[10] Garvey (2001).
[11] *Id.* at p. 2.

high-value added, time-sensitive cargo. These carriers guarantee the timely delivery of these vast volumes of shipments, ranging from same-day delivery to 72 h after pick-up, virtually anywhere in the world. They operate in 220 countries and territories.

The conference noted that a real risk in the area of cargo and mail security would arise when an express delivery carrier experiences a technical problem in an aircraft and is forced to transfer cargo to a passenger carrier, in which instance strict supply chain standards should be adhered to so that the risk in the transfer of cargo could be obviated.

Participants agreed that it was essential that solid standards and mutual recognition programmes be in place in order to make sure that States all along an air cargo supply chain satisfy themselves that air cargo is secure, and so let it flow unimpeded. Such standards and recommended practices should allow for the speedy transit and transhipment of legitimate air cargo worldwide, through any combination of air routes and transit or transhipment points.

These two examples contrast with the mere lip service given by ICAO to the economic regulation of air transport. The reason is clear.

1.2   The Solution

Unlike air navigation and security, which have to be globally harmonized and standardized[12] no matter what the airlines involved are, and what the States are, and which have no bearing on market economics or competition, the economics of air transport has necessarily two key players which, if they do not go hand in hand, would be isolated and lost. The two players are States and commercial air carriers, both of whom should drive their efforts towards one goal—to meet the needs of the people for regular, economic and efficient air transport. Although ICAO espouses liberalization of air transport services, it does not, and indeed has not the teeth to enforce liberalization on States. States therefore could enforce their interests and those protective interests of their national carriers in ensuring their market share, to the exclusion of competition. IATA,[13] which is the association of the airlines on the

---

[12] Harmonization in this context means consistence in the various national regional and global rules, regulations procedures and practices. Standardization is conformance and compliance.

[13] The International Air Transport Association, an association of air carriers, was formed in 1919 as the International Air Traffic Association. Encapsulated in IATA's overall mission are seven core objectives: to promote safe, reliable and secure air services; to achieve recognition of the importance of a healthy air transport industry to worldwide social and economic development; to assist the air transport industry in achieving adequate levels of profitability; to provide high quality, value for money, industry-required products and services that meet the needs of the customer; to develop cost effective, environmentally-friendly standards and procedures to facilitate the operation of international air transport; to identify and articulate common industry positions and support the resolution of key industry issues; and to provide a working environment which attracts, retains and develops committed employees.

other hand, looks after the interests of the carriers, but mostly their protection from excessive charges, rates and restrictive rules on the operation of air services.

The centrifugal point for air transport economics is competition. Both ICAO and IATA should work towards ensuring competition as a common goal of air transport. This would not only satisfy ICAO's aim under the Chicago Convention to "meet the needs of the people of the world for safe, regular, economical and efficient air transport" but also fulfil IATA's objective of moving towards "a safe, secure and sustainable air transport industry that connects and enriches our world".

First off, ATConf/6 should have been a joint ICAO/IATA Conference where both parties should have faced the aviation community on a common front as discussed above. This has never happened before. The primary product, which is the air transport product, comes first, and airports, air navigation services providers and other service providers come second. The theme of ATConf/6 should have been *"Connectivity through Competition"*, whereas the theme of the Conference was *"Sustainability of Air Transport"* and the objectives of the Conference were to develop guidelines and an action plan for a global regulatory framework. Taken separately both the theme and objectives reiterate the clichés of the past decades. Is it only now that we are thinking of "sustainability of air transport"? And what exactly does it mean/Has air transport been unsustainable? And if so, why have we not had a sustained air transport system for the past 67 years? And what about the objective of "developing guidelines"? Haven't we had enough guidelines? What magical guidelines are these? And how effective are guidelines? How would a "global regulatory framework" work, when States and the airlines would pursue their own policies?

Unfortunately, once again, ICAO missed the bus and went through the ritual of convening a meeting that would come up with "guidelines". What if the Conference had an action plan, which the States bought into at the Conference, which forced their hands to achieve connectivity through competition? The European Regional Seminar of the European Civil Aviation Commission (ECAC) on 14 November 2012 suggested the adoption by ICAO of basic principles for open and fair competition and related policy instruments. At the time of writing, the most recent development on this issue was the conclusions adopted by the Council of the European Union (EU) on 20 December 2012. These conclusions recognized that the EU aviation sector and EU airlines are facing difficult challenges in a highly competitive global aviation market and, inter alia, concluded that ATConf/6 presents an opportunity for pursuing a "level playing field" through a new framework for fair competition.

Connectivity[14] is a property of a network and can be defined in such a way as to constitute an indicator of the network's concentration. Therefore, connectivity is the ability of a network to move a passenger from one point to another with the lowest possible number of connections and without an increase in fare, focusing on,

---

[14] For purposes of this book, connectivity refers to optimizing airline connections through market access.

from a commercial perspective, minimum connecting times with maximum facilitation ultimately resulting in benefits to air transport users.

This broad definition of air transport connectivity illustrates that there are several factors which enable connectivity, from availability of air transport services and airline practices, to security and facilitation procedures. In past years, the pace of liberalization has been affected by the global economic recession, which has led to a reversion toward trade protectionism in many countries. Air transport has not escaped this setback, even though in practice liberalization could be a stimulus to recovery. As of today, the vast majority of arrangements are under this type of bilateral Air Services Agreement, and the primary focus is on removing national air carrier ownership and control provisions from air services agreements and agreeing on acceptance of ownership and control provisions based on principal place of business in one or more of the participating States. Therefore, the key challenge is linked to bilateral restrictions which limit the availability of services for the air transport user. According to the industry, airline liberalization can further increase demand and ensure that the services, which are providing increased connectivity, are sustainable over the long-term. It provides the commercial freedom necessary for airlines to adjust capacity appropriately to meet changes in air travel demand. By way of example, the growth in air services between Poland and the United Kingdom (UK) since 2003 resulted in a gross domestic product (GDP) increase of 27 % for Poland, whereas the increase in the already well-served UK was a much smaller 0.5 %. These changes provide an estimated long-term boost to Poland's GDP of US$634 million per annum. The UK also benefited, with an estimated boost to its GDP of US$45 million per annum.

Network carriers generally offer scheduled flights to major domestic and international cities while also serving smaller cities; the carriers normally concentrate most of their operations in a limited number of hub cities, serving most other destinations in the network by providing one-stop or connecting service through the hubs. An efficient utilization of the hub allows airlines to offer better connectivity. The "hub and spoke" model is a system which enhances efficiency in transportation by greatly simplifying a network of routes. Many airlines supplement the "hub and spoke" model with codeshares, partner flights, or a small commuter airline. It should be noted that the way in which airlines price tickets can also impact connectivity, notably in the case of transit by flight stage; if a trip is sold by flight stage as opposed to origin to destination.

The lack of leadership in the economic regulation of air transport and the various "open skies" regimes that have sprung up along with mergers and acquisitions of airlines resulting in airline alliances and other commercial arrangements, is symptomatic of a crisis in law that has to be addressed. *Ex facie*, this would call for a three dimensional approach where a triage of institutions, i.e. the two parties to an agreement on market access would be refereed and regulated by a third party that would guarantee fairness of trade ensure that the conduct of trade in air transport accords with global practices of fair trade from a consumer perspective, would act as key players. This would initially mean that States veer from the parochial dogma of absolute state sovereignty and embrace the concept of sovereignty that is consistent with globalization and accepted trade practices. In its Report to the General Assembly, the International Law Commission recommended a draft provision which required:

> Every State has the duty to conduct its relations with other States in accordance with international law and with the principle that the sovereignty of each State is subject to the supremacy of international law.[15]

This principle, which forms a cornerstone of international conduct by States, provides the basis for strengthening international comity and regulating the conduct of States both internally—within their territories—and externally, towards other States. States are effectively precluded by this principle of pursuing their own interests untrammelled and with disregard to principles established by international law. State Sovereignty thus connotes a responsibility rather than an absolute right to do as a State deems fit within its own territory. The conduct of trade in air transport should be determined by an international political process or a "central market place". Supiot uses the metaphor of the medieval *Marktplatz* of Brussels:

> Its architectural magnificence is imbued with institutional meaning. This ancient market place is bounded by the headquarters of the institutions which endured the smooth operations of the market. The Town Hall housed the municipal authority that saw to the fairness of trade through inspection of weights and measures, while the buildings of various trades (e.g. butchers, bakers, brewers) housed the guilds that upheld the status and quality of labour, without which there could have been nothing valuable to trade. These various buildings also marked out the boundaries of the commercial sphere. If one left the market place say, to go to the courthouse or to the Royal Palace, a different set of rules applied. Indeed if the law of the market had extended to judges or to political leaders, their decisions would have been up for sale, the city would have been corrupt, and honest traders would have been unable to carry on their business there freely.[16]

Supiot makes a good point which is particularly applicable to the air transport analogy. The lack of central laws and regulations in market access that would otherwise ensure fairness of competition and worn out perception of sovereignty has made States control market access through a certain parochial protectionism, opening the door for various "deals" in the market.

The problem is aggravated by the fact that in modern parlance the various freedoms associated with free trade such as the freedom of establishment; the freedom to provide services; and the freedom to move capital and goods all encourage investors to go "forum shopping" so that they could opt to establish themselves in the jurisdiction that is most conducive to their interests. This allows investors to bypass the jurisdiction that they would be subject to if they were to

---

[15] *Report of the International Law Commission to the General Assembly on the Work of the 1st Session, A/CN.4/13*, 9 June 1949, at 21.

[16] Supiot (2010), pp. 151–162 at 152.

establish their businesses in their States of nationality and seek less constraining jurisdictions of their choice.[17]

The issue for international air transport is "what is the central market place?" The immediate answer which comes to mind, if air transport were to be treated as a trade, is the World Trade Organization (WTO). One of the most contentious issues in the world of commercial air transport today is the question as to whether the industry should embrace the trade in services regime of the World Trade Organization in preference to the currently restrictive system of the bilateral air transport agreement which entitles States to refuse permission to air carriers who apply to operate commercial air services to and from their territories. The current system is based on national treatment, where States could apply different conditions to carriers operating air services into their territories based on capacity and demand for travel at prices that are deemed acceptable to the State concerned and the travelling public. Corollaries to the current system of bilateral air transport agreement are traditionally seeped in norms requiring capacity to be primarily provided for traffic to and from the two States parties to the agreement, on the basis of "fair and equal opportunity". Under this system, airlines are usually expected to be substantially owned and effectively controlled by a minimum percentage of the Contracting party. This affords a State the opportunity to protect the interest of its own carriers.

Air transport affects world trade in two ways: as a service by itself, directly transporting persons and freight; and as a service feeding other areas of trade mainly involving tourism and hospitality. However, it cannot be doubted that air transport affects overall economic activities of business, particularly involving cross-border trade. It is also incontrovertible that the world needs an efficient and effective air transport industry if the dual functions of the air transport service were to be sustained over time to cater to the rapidly growing demand for carriage by air of persons and goods. In order to achieve this objective, the air transport industry must be liberalized to the extent that it remains unfettered by commercial constraints. Yet, unlike most other modes of transport, air transport remains rather rigidly controlled by the need for agreements of States to permit carriers of other States into their territories as well as established percentages of national ownership of air carriers which stifle foreign investment in national jurisdictions. While at least one commentator has categorically stated that the trend towards a very liberal open skies international regime is unstoppable,[18] which implicitly gives the industry the assurance that the problem would solve itself in the years to come, others have vigorously advocated that, as a panacea to the problem of rigid regulation, market access in air transport should be in the domain of a liberalized international

---

[17] See *International Transport Workers' Federation and Finnish Seamens' Union* v. *Viking Line ABP and OU Viking Line Easti*, ECR 2007, 1-10779 (Case C-438/05, 6 December 2007). Also, *Centros Ltd.* v. *Erhvervs-og Selskabsstyrelsen*, ECR, 1-1459case no. C-212/97, 9 March 1999. For an air transport analogy see Abeyratne (2004), pp. 585–601.

[18] Doganis (2001) at 11.

regime. While the former view cannot be disputed, the latter approach brings to bear the compelling need to address the issue squarely, both in terms of whether the desirable approach would be to bring the industry from the current bilateral structure of air services negotiations into a more generalized regime and if so, what the modalities of such an exercise might entail. As to the former, it is largely a matter of political will. The latter would need some discussion on the legalities involved.

1.3 Competition

Competition strategy is composed of calculated business acts which deliberately choose a different set of activities to deliver a unique mix of value.[19] National competitiveness is one of the most critical drivers of successful government and industry in every nation. Yet for all the discussion, debate, and writing on the topic, there is still no persuasive theory to explain national competitiveness. What is more, there is not even an accepted definition of the term "competitiveness" as applied to a nation. While the notion of a competitive company is clear, the notion of a competitive nation is not. The deliberations of ATConf/6 clearly brought to bear this point and implicitly called upon States to revisit their own strategies with regard to the air transport industry on the basis that the most important feature of a competitive nation is its decisive characteristic that allows its companies to create and sustain competitive advantage in particular fields—the search is for the competitive advantage of nations Of particular concern are the determinants of international success in technology and skill-intensive segments and industries, which underpin high and rising productivity.

Classical theory supports the principle that the success of nations in particular industries based on so-called factors of production such as land, labor, and natural resources is based on the fact that nations gain factor-based comparative advantage in industries that make intensive use of the factors they possess in abundance. Classical theory, however, has been overshadowed in advanced industries and economies by the globalization of competition and the power of technology.

Any new approach must recognize that in modern international competition, companies compete with global strategies involving not only trade but also foreign investment. What a new theory must explain is why a nation provides a favorable home base for companies that compete internationally. The home base is the nation in which the essential competitive advantages of the enterprise are created and sustained. It is where a company's strategy is set, where the core product and process technology is created and maintained, and where the most productive jobs and most advanced skills are located. The presence of the home base in a nation has

---

[19] Porter (1998) at p. 45.

the greatest positive influence on other linked domestic industries and leads to other benefits in the nation's economy. While the ownership of the company is often concentrated at the home base, the nationality of shareholders is secondary.

A new theory for air transport must seemingly move beyond comparative advantage to the competitive advantage of a nation. It must reflect a rich conception of competition that includes segmented markets, differentiated products, technology differences, and economies of scale. A new theory must go beyond cost and explain why companies from some nations are better than others at creating advantages based on quality, features, and new product innovation. A new theory must begin from the premise that competition is dynamic and evolving; it must answer the questions: Why do some companies based in some nations innovate more than others? The most important feature of a competitive nation is its decisive characteristic that allows its companies to create and sustain competitive advantage in particular fields—the search is for the competitive advantage of nations. Of particular concern are the determinants of international success in technology and skill-intensive segments and industries, which underpin high and rising productivity.

Any new approach must recognize that in modern international competition, companies compete with global strategies involving not only trade but also foreign investment. What a new theory must explain is why a nation provides a favorable home base for companies that compete internationally. The home base is the nation in which the essential competitive advantages of the enterprise are created and sustained. It is where a company's strategy is set, where the core product and process technology is created and maintained, and where the most productive jobs and most advanced skills are located. The presence of the home base in a nation has the greatest positive influence on other linked domestic industries and leads to other benefits in the nation's economy. While the ownership of the company is often concentrated at the home base, the nationality of shareholders is secondary.

In the continuing debate over the competitiveness of nations, no topic engenders more argument or creates less understanding than the role of the government. Is government an essential helper or supporter of industry, employing a host of policies to contribute directly to the competitive performance of strategic or target industries? Or is it the "free market" view that the operation of the economy should be left to the workings of the invisible hand.

Both views are seemingly incorrect. Either, followed to its logical outcome, would lead to the permanent erosion of a country's competitive capabilities. On one hand, advocates of government help for industry frequently propose policies that would actually hurt companies in the long run and only create the demand for more helping. On the other hand, advocates of a diminished government presence ignore the legitimate role that government plays in shaping the context and institutional structure surrounding companies and in creating an environment that stimulates companies to gain competitive advantage.

Government's proper role is as a catalyst and challenger; it is to encourage—or even push—companies to raise their aspirations and move to higher levels of competitive performance, even though this process may be inherently unpleasant and difficult. In liberalizing international air services, States have opted for greater reliance on competitive forces and, accordingly, have a greater need to address potential anti-competitive abuse or collusion than in the past era of tight government regulation of routes, frequencies, and fares. A fundamental issue is the question of the tools best suited to address this concern. Options include a continued reliance on mechanisms within air services agreements, the application to international air services of national competition laws or policies of general applicability, or some combination of the two.

Although there is presently no consensus on a single approved approach, certain conclusions may be drawn. The first is that, in a liberalized market, the distortive effect of unilateral government assistance known as "state aids" has been broadly recognized. Such assistance is largely outside the scope of national competition laws, which address the actions of private sector competitors, and is thus more amenable to regulation through provisions in ASAs. This point is underscored by the inclusion in a growing number of liberalized air services agreements of non-exhaustive enumerations of conditions likely to adversely affect a fair and competitive environment, such as capital injections, cross subsidization, grants, guarantees, government ownership, tax relief or tax exemption and protection against bankruptcy or insurance by a government entity. It should be noted that there is also control of state aid at the European Union level in order to avoid distortions within the Single Market.

A second conclusion is that States must exercise care in applying their national competition laws and policies to international air services. With increased globalization and the adoption of market economy principles in aviation, it is not surprising that States would respond by seeking to apply their competition regulations to the sector. However, national competition laws, notably those governing mergers or alliances, may conflict with each other. Moreover, certain States do not have competition laws. Furthermore, the traditional approach in many bilateral agreements favouring airline cooperation on issues like capacity and pricing is squarely at odds with competition laws that strictly prohibit price-fixing, market division and other collusive practices by market competitors.

In cases where national competition laws or policies are applied to international air transport, States should give due consideration to the concerns of other States involved. Cooperation is needed between States, and especially between competition authorities. At a minimum, such cooperation should aim at avoiding outright conflicts of legal obligations placed upon airlines. Although wide-ranging legal harmonization remains a distant objective, consultation and information-sharing between competition authorities can foster better understanding and, as has been demonstrated in the review of certain airline alliance agreements, greater compatibility in competition law analyses and remedies. For States that do not have competition laws, additional ICAO guidance could be developed for inclusion in air services agreements to ensure that airlines operate in a framework covering the basic competition affecting them.

A theme often sounded in the discussion of international air transport competition is the need for a "level playing field". On an abstract level, there is broad acceptance of the principle that a fair and equal opportunity is required to allow airlines to succeed and grow in the liberalized global market. In fact, some air services agreements refer explicitly to the principle of a level playing field by noting that "where there is not a level competitive playing field for airlines, potential benefits deriving from competitive air services may not be realised". However, it must be recognized that there is currently no commonly accepted definition of the conditions constituting a "level playing field". It is unlikely that consensus on a comprehensive definition can be achieved at this time, given the widely different circumstances of States and their aviation sectors, including such fundamental issues as state ownership, policies on maintenance of national air carriers and airport development, and widely divergent State policies on taxation, labour regulation, bankruptcy, and health insurance.

IATA has some words of wisdom on this subject:

> In the absence of a fully liberalized regime, alliances, code sharing and joint ventures, subject to strict compliance with applicable competition laws, have become a feature of the industry, acting as a substitute for mergers and acquisitions. They represent a second-best alternative for airlines to offer a consistent travel experience to customers and benefit from economies of scale where possible, given the current restrictive bilateral air service agreements. However, the lack of coordination among competition authorities around the world presents a challenge to the establishment of such arrangements. Competition authorities are encouraged to take a coordinated approach to approval and/or enforcement of existing laws when examining such cross-border joint ventures.[20]

1.4 The ICAO-IATA Synergy

ICAO's mission statement speaks of capacity building through cooperation among States and stakeholders. The word "stakeholder" once avoided by the ICAO Council (in the last decade), has resurfaced. It is a term used mostly in project management where a stakeholder is defined as "a person or group of people who can affect or be affected by a given project". Stakeholders can be individuals working on a project, groups of people or organizations, or even segments of a population. A stakeholder may be actively involved in a project's work, affected by the project's outcome, or in a position to affect the project's success. Stakeholders can be an internal part of a project's organization, or external, such as customers, creditors, unions, or members of a community.

ICAO's ongoing work in the key areas of safety, security, environmental protection, and now the development of air transport, is by no means a project. This being eliminated, we could turn to the general definition of a stakeholder, which is "A person, group or organization that has interest or concern in an organization" Stakeholders can affect or be affected by the organization's actions, objectives and policies.[21]

---

[20] IATA Position Paper on Liberalization at http://www.iata.org/policy/Pages/liberalization.aspx.
[21] http://www.businessdictionary.com/definition/stakeholder.html.

The question is, is IATA a stakeholder in ICAO's activities? If current reality were to be applied, the answer has to be in the negative as IATA is a mere observer in ICAO's Air Transport Committee—a subsidiary body of the ICAO Council addressing economic issues. IATA represents airlines, which are not even remotely connected to ICAO and its work. ICAO has signed numerous arrangements of cooperation with the so called "stakeholders" including Airports Council International (ACI), Civil Air Navigation Services Organization (CANSO) and IATA, none of which directly addresses ICAO's aim to meet the needs of the people of the world for economical air services. This startling inconsistency continues to be perpetuated, where IATA repeatedly submits working papers at ICAO conferences (among which was ATConf/6) as an observer.

IATA's mission is to represent, lead and serve the airline industry. How would this mission be accomplished if it does not have an influential voice at ICAO? For example, IATA claims that liberalization of air transport is the right path to follow, but a fully liberalized regime will take years to achieve. In the interim, there are short-term options that governments can pursue which will enhance connectivity and maximize aviation's contribution to the global economy and society. A position paper issued by IATA on liberalization states that improved connectivity brought about by aviation opens up new trade markets, and encourages innovation, competition and more choice. Economic studies show that a 10 % increase in global connectivity relative to GDP would see a 0.07 % per annum increase in long-run GDP1. Aviation also provides significant social benefits; uniting people and facilitating cultural understanding as well as exchange of ideas.

In the same vein IATA claims: "IATA supports ICAO's leadership role in the area of liberalization and acknowledges the considerable ICAO policy and guidance material that has been developed in the areas of market access and foreign ownership and control. ICAO is the right forum to lead air transport to a more liberalized environment". The trick here is to reconcile the words "leadership" and "policy and guidance material". If ICAO leads in the area of liberalization, why is it restricted to policy and guidance material?

The only way forward in this dichotomy would be to define ICAO's "leadership" within the meaning and purpose of fostering the development of air transport through meeting the needs of the travelling public. This statement should be read proactively, with emphasis on the latter element rather than the former in which case IATA becomes not just an observer but an active partner. The structured way to approach this problem would be to firstly identify exactly what ICAO's role in the economic aspects of air transport. If the central point is the consumer, which indeed it should be, ICAO's leadership should be conjoined with that of IATA where the airlines should be provided specific directives to maximize connectivity, thereby giving the consumer regular, economic and efficient air transport, as the Chicago Convention requires.

Ideally, there should have been a paper presented by the ICAO Secretariat at ATConf/6 on "ICAO's Future Role on the Economic Aspects of Air Transport" which should have cleared the air and effectively precluded the aviation community from maundering in circumlocution on the subject. That paper should have particularly identified the relationship between ICAO and IATA in more proactive terms than the lip service given at present. In this context, ICAO should follow the United Nations and its connectivity at the present time with civil society. Secretary General Ban Ki-Moon said:

> Our times demand a new definition of leadership – global leadership. They demand a new constellation of international cooperation – governments, civil society and the private sector, working together for a collective global good[22]:

The United Nations is both a participant in and a witness to an increasingly global civil society. More and more, non-governmental organizations (NGOs) and other civil society organizations (CSOs) are UN system partners and valuable UN links to civil society. CSOs play a key role at major United Nations Conferences and are indispensable partners for UN efforts at the country level. NGOs are consulted on UN policy and programme matters. The UN organizes and hosts, on a regular basis, briefings, meetings and conferences for NGO representatives who are accredited to UN offices, programmes and agencies.

Pursuant to the United Nations Secretary General Kofi Annan's initiative in 2009 "Strengthening of the United Nations: an agenda for further change",[23] in February 2003, he appointed the Panel of Eminent Persons on United Nations–Civil Society Relations. Their Report can be taken as an analogy for ICAO's way forward. The Panel believed that constructively engaging with civil society is a necessity for the United Nations, not an option. This engagement is essential to enable the Organization to better identify global priorities and to mobilize all resources to deal with the task at hand. We also see this opening up of the United Nations to a plurality of constituencies and actors not as a threat to Governments, but as a powerful way to reinvigorate the intergovernmental process itself.

The Panel's Report underscored the point that the world stands today at a very delicate juncture. The United Nations needs the support of civil society more than ever before. But it will not get that support unless it is seen as championing reforms in global governance that civil society is calling for. Over the years, the relationship of the United Nations to civil society has strengthened and multiplied.

The Report underscored that the United Nations will be more effective if it reached out beyond its constituency of central Governments and enhanced dialogue and cooperation with civil society. The driver was global change and the attributes of many civil society organizations, which brought to bear the fact that an enhanced engagement could help the United Nations do a better job, further its global goals, become more attuned and responsive to citizens' concerns and enlist greater public support. Admittedly, there were trade-offs. The most powerful case for collaboration with civic society was that the unique role of the United Nations as an intergovernmental forum is vitally important and must be protected at all costs. Today's challenges require the United Nations to be more than just an

---

[22] Secretary-General Ban Ki-moon's Speech at World Economic Forum Davos, Switzerland (29 January 2009).

[23] A/57/387 and Corr.1.

intergovernmental forum; it must engage others too. To do so risks putting more pressure on the Organization's meeting rooms and agendas, which are becoming ever more crowded; this calls for more selective and not just increased engagement.

One cannot argue against the logic of the Report and its applicability to ICAO. Furthermore, the principles of the Report that support the thrust of opening out the United Nations goes for ICAO too. For instance, the following could apply to a synergy between ICAO and IATA:

- Become an outward-looking organization. The changing nature of multilateralism to mean multiple constituencies entails ICAO giving more emphasis to convening and facilitating rather than "doing" and putting the issues, not the institution, at the centre.
- Embrace a plurality of constituencies. Many actors may be relevant to an issue, and new partnerships are needed to tackle global challenges.
- Connect the local with the global. The deliberative and operational spheres of ICAO s are separated by a wide gulf, which hampers both in all areas from development to security. A closer two-way connection between them is imperative so that local operational work truly helps to realize the global goals and that global deliberations are informed by local reality. Civil society is vital for both directions. Hence the country level should be the starting point for engagement in both the operational and deliberative processes.

ICAO's 191 member States—mostly the developing States—look up to the Organization for moral leadership. ICAO should use this moral leadership to forge coordinated approaches to civil society, to encourage Governments to provide a more enabling and cooperative environment for civil society and to foster debate about reforms of global governance, including deeper roles for civil society. This should emphasize principles of constituency engagement, partnership, transparency and inclusion, with a special emphasis on those who are normally unrepresented, such as the consumer of air transport.

Other United Nations specialized agencies have preceded ICAO in this regard. In the context of the World Health Organization (WHO), the first World Health Assembly adopted Principles to govern relations between WHO and nongovernmental organizations (NGOs) in furtherance of Article 71 of the WHO Constitution, which provides for arrangements to be made for ". . . consultation and co-operation with non-governmental international organizations and, with the consent of the Government concerned with national organizations, governmental or non-governmental." For the purposes of the policy, nongovernmental organizations (NGOs) includes such organizational forms as civil society organizations, associations of professions, industries, patients, foundations, service providers. WHO has a Partnerships Unit for this purpose. enabling informal relations to flourish between NGOs and all levels of WHO, from the offices of the WHO representatives at the country level, to regional offices, and the headquarters in Geneva. They form the majority of WHO/NGO relations and are pursued freely without the need for central reporting. The Principles also set out the terms for the admission of international NGOs into official relations with WHO, that is, formal relations.

With regard to the International Telecommunications Union (ITU), the issue of relations with development NGOs goes back some time. Currently, like ICAO, the ITU has no specific mandate to include civil society or NGOs in its activities. ITU members in theory comprise governments only, but in the last decade or two it has gone to great lengths to ensure that the private sector can also have a full say in all but its formal internal structures. However, development NGOs remain officially ignored and are simply not recognised as having anything to contribute in the area of international telecommunications.

As for the United Nations Educational, Scientific and Cultural Organization (UNESCO), it may establish official partnerships with international, regional, national or local non-governmental organizations (NGOs). Any organization may qualify as a non-governmental organization which may become a partner of UNESCO provided that it has not been established by intergovernmental agreement, or by a government and that its purposes, functions, structure and operation are non-governmental, democratic and non-profit-making in character. The international or regional NGO must form, through its regular active membership (consisting of institutions and/or individuals), a community linked by a desire to pursue, in a significant number of countries or regions, the objectives for which it was established.

The World Meteorological Organization (WMO) is a UN specialized agency, tasked to coordinate observations on the state of the global weather and climate systems and support the communication of data on weather and climate variability and change. As weather, climate and climate variability and change have a significant and increasing effect on global economies, WMO seeks to further engage the business sector to expand the observing and applications networks and ensure the provision of better and more user-friendly data and information services. WMO seeks new partnerships with the private sector that can support our work at all levels. The desired partnerships can take the form of involving the business sector in our programmes, fundraising support or contributions in-kind.

ICAO is an inter-governmental organization and, being that, can have only States as members. Therefore it goes without saying that IATA cannot be a member of ICAO. However, following the United Nations approach and the path followed by many specialized agencies, a closer partnership with IATA will serve ICAO's objectives much more effectively and efficiently. To sum up with the words of Tony Tyler, IATA Director General:

> Partnership is critical. It is no accident that ICAO and IATA both call Montreal home. Working together is part of the DNA of both Organizations. For over six decades we have worked together to develop harmonized standards, recommended practices and guidance in many areas...I see broader regulatory issues taking on a bigger role in the ICAO IATA joint work...[24]

---

[24] *ICAO Journal*, Vol. 67, No. 6. At 11.

1.5   Governance

Lastly, a word about governance. Governance is about principles and responsibilities that achieve a strategic direction. For there to be effective governance, there must essentially be strategic direction, and that is what ICAO lacks in the field of economic regulation. ICAO needs a unified governance system through reorganization. Currently, ICAO has a Headquarters and seven regional offices. These regional offices, although nominally and theoretically controlled by the Headquarters, operate quite autonomously. This system is analogous to a country establishing regional police stations which are not totally controlled by the central government. This dysfunctional structure leaves some States without the services that they require, particularly access to information.

ICAO has exclusively concentrated on technical issues in the regions. There are no air transport officers anymore. The Secretary General of ICAO, at an informal briefing on regional activities of the Organization delivered to the Council on 28 April 2010 mapped out certain facts which shows the lack of involvement in economic issues pertaining to air transport in the regions and the emphasis on technical issues. The Secretary General stated:

> The regional offices were set up almost concurrently with the coming into being of ICAO. They were established for the purpose of ensuring safety of aviation in the regions including but not limited to maintaining liaison with States on the regional plans, conducting missions to States, preparation of special studies of interest to States in the region; identifying and addressing deficiencies, and maintaining diplomatic relations with States.

By his own admission, the Secretary General recognized that over the past several years, the Regional Offices and their functions have enjoyed some exclusivity with regard to Headquarters and at present, there isn't enough cooperation between the regional offices and Headquarters. In the Europe and Middle East geographical area (i.e. Paris and Cairo), the Regional Offices participate in a limited way in technical cooperation activities (note the absolute lack of reference to economic activities). The Secretary General went on to say that the small participation of the Paris Office was understood, considering the small number of technical cooperation projects in Europe. In the Middle East region, where there are a number of important technical cooperation projects, the participation of the Cairo Office is more significant.

His main concern at that time was that there was no Technical Cooperation Officer in the Office, thus hindering the full potential of collaboration between the Regional Office and TCB. On the other hand, the Cairo Regional Office is kept well informed of technical cooperation projects in the region, particularly with regards to regional projects. In fact, the Regional Office in Cairo participates actively in the Board meetings of the Middle East Regional Monitoring Agency project.

In the Americas geographical area (i.e. Mexico and Lima), the Regional Offices participate actively in a number of technical cooperation activities. With regards to regional projects, meetings, seminars or conferences concerning technical cooperation, the Regional Offices are always invited to participate and in many instances have a lead role in the organization of such activities. For example, the Regional Office in Mexico administers, in close collaboration with TCB, regional projects such as the CAPSCA, as well as coordinating important meetings such as the NACC/DCA steering committee meetings in the CAR sub-region. It was nonetheless a matter of concern that neither of the Regional Offices in the Americas benefitted from the support of a Technical Cooperation Officer, which could enhance ICAO's mission in the region, particularly with respect to technical cooperation project development.

The Secretary General also noted that in Africa (i.e. Dakar and Nairobi), the participation of the Regional Offices in technical cooperation activities was inconsistent, most notably in Dakar, which does not have a Technical Cooperation Officer. In Nairobi however, a Technical Cooperation Officer was appointed to assist in the promotion and management of technical cooperation activities.

This blatant lack of recognition of economic regulatory support and direction to a region that most needs it from ICAO causes some confusion in the context of the current needs of the 57 African States. The air transport industry concerning Africa, which includes the African air transport industry and the international air carriers operating services into Africa, is mostly driven by the trade generated by natural resources available in the African Continent. In the earliest years of the current century, after the cessation of operations of both Sabena and Swissair, there was a lacuna that other carriers had to fill in order to provide capacity needed for the African trade machine to run smoothly. The example of quick action by Air France, partially concerning the Congolese Kinshasa market, and the provision of much needed capacity, is cited by economic experts as typical of the African market.[25] The injector of capacity by Air France to a former colony of France reflects well the air transport considerations of the Continent, where, irrespective of the ravages of war, the mineral resources available, the need to build infrastructure and the steady influx of businessmen to the Kinshasa area overshadowed civil unrest, calling for a steady flow of air transport.

When considered sub-regionally, Africa shows distinct characteristics in each area. These characteristics are essentially response based, showing the connections of Africa to the rest of the world. For example, North Africa demands and sustains a developed air transport system and infrastructure, in view of its close geographic proximity to Europe and the Mediterranean World. Airlines such as Egypt Air and Royal Air Maroc, including others in the area, have moved towards regional point to point operations as well as international point to point services, showing a distinct move toward the maximizing of capacity, services and profitability.

In terms of Western and Central Africa, which was most affected by the liquidation of Sabena and Air Afrique, and where some East African States felt the impact of the lacuna created by the cessation of services by these carriers, a revival is visible. While Air France has increased services and supplemented the

---

[25] African Air Transport Capacity, Africa is Ready for Action, *The Avmark Aviation Economist*, October 2002 at p. 17.

provision of much of the capacity cost, new local carriers are starting up to grab the opportunity. Examples are Africa Air and Africa One, both of Uganda who have the ambition of introducing flights from Entebbe to some points in Africa and the Middle East. Particular mention must be made of the burgeoning air transport potential of Dakar, Senegal, which is also the home to a regional office of ICAO as well as to Senegal International, a carrier with significantly great potential that could establish Dakar as a hub for international and intra-continental traffic.

East Africa is also surging ahead, a good example of which is Ethiopia which is progressively strengthening its aviation infrastructure and regulatory governance. As a result, Ethiopian Airlines has developed a substantial route network, converting Addis Ababa into a busy hub. The airline's main focus is on international markets with particular emphasis on North America. Kenya, on the other hand is fortunate to be threatened by competition, particularly from such States as Tanzania in the area of tourism. This has impelled Kenya Airways to aggressively buttress its presence in the region and the government of Kenya to strengthen its already powerful tourism base. Efforts of Kenya Airways to modernize its fleet with Boeing 737-700 and 777-200 ER aircraft portend well for competition within the region.

The remaining Southern region, of which South Africa is the major player, boasts of South African Airways, the largest and most reputed carrier in the African Continent. The airline leads the Alliance and has regional subsidiaries, SA Express and SA Airline, to supplement the parent SAA's international operations. One of the more desirable features of having such a large and efficient carrier in the African Continent is more involvement on a regional basis of the carrier in partnerships which may develop later into alliances with carriers of other regions. This has yet to happen.

Airports and air navigation services form an integral part of the African aviation industry and cannot be ignored in balancing the air transport equation in the Continent. Although Africa accounts approximately for 2 % of the global air traffic, the compelling need for an air transport network in the Continent and the inevitable dependence placed on air transport overshadows statistical determination. In general terms, surface transport and roadways throughout most of Africa are not adequately developed to ensure smooth communications—a fact which is brought to bear by the practical necessity in some instances of travel between two neighbouring African States via a European point such as Paris by air. However, most African airports have not been improved upon or extended since their construction in the 1950s and 1960s and therefore a sustained airport and air navigation infrastructure development programme for Africa is much needed, both from technological and economic perspectives. It must be recognized that African States have shown a conscious awareness of this need, which is borne out by such examples as the involvement in Kilimanjaro Airport by Kilimanjaro Airports Development Company (KADCO—a private company). Other airports which are open to private management are Algiers Airport, Djibouti Airport and Bamako Airport.

One of the watershed events of African civil aviation history occurred in 1961 when ten African nations[26] signed the Treaty on Air Transport in Africa.[27] Popularly known as the Yaoundé Treaty, it has its roots in Articles 77 and 79 of the Chicago Convention of 1944[28] which provides for the setting up by two or more States of joint or international operating organizations and for the participation of the States in these organizations. Based on these principles, the Yaounde Treaty established Air Afrique to operate on behalf of its contracting States international services between their territories and from their territories to non-contracting States' territories and also domestic air services within the territories of the contracting States.[29] These services were operated by the airline, until it ceased to operate, consequent to the negotiations between contracting States carried out through a body named Comite Multinational de Negociation des Etats Signataires du Traité de Yaounde (CMN). At the present time, contracting States to the Yaounde Treaty include States of Western and Central African sub-regions.

The second major event in African civil aviation history occurred in 1988 when African civil aviation ministers gathered in Yamoussoukro in the Republic of Côte d'Ivoire on 6 and 7 October and signed the Yamoussoukro Declaration on a New African Air Transport Policy.[30] This declaration was the result of a collective consensus in Africa that African nations must, *inter alia*, prepare for the effects of deregulation in the United States on other countries and the potential adverse effects on African airlines of the air transport liberalization policies of Western Europe, especially the application by EEC of the Treaty of Rome to air transport services and the creation of a single internal European market by 1993. The Declaration also responds to the fact that many aircraft owned by African airlines are obsolete and thus in need of replacement at great cost, particularly with regard to the need for African airlines to comply with Chapters 2 and 3 of the Annex 16 to the Chicago Convention concerning aircraft noise. Another compelling issue addressed at Yamoussoukro and incorporated into the Declaration concerned the liberal exchange of air traffic rights by African States and the need for African airlines to market their product competitively through an unbiased computer reservation system.

The Yamoussoukro Declaration committed African States both individually and collectively, to achieve the total integration of their airlines under the above policies within a period of 8 years. The 8 years was divided into three phases of 2 years, 3 years and 3 years respectively. A review of the Declaration, conducted by

[26] Central African Republic, Republic of the Congo, Republic of the Ivory Coast, Republic of Gabon, Republic of Dahomey, Republic of Upper Volta, Islamic Republic of Mauritania, Republic of Niger, Republic of Senegal and the Republic of Chad.

[27] See ICAO Circular 98-AT/19, 1970 for text of the Treaty.

[28] See, Convention on International Civil Aviation, signed at Chicago on 7 December 1944. Cited at supra, note 1.

[29] *Id.* Article 3.

[30] See Report of the African Civil Aviation Commission, Eleventh Plenary Session (AFCAC/11), Blantyre, 22–31 May 1989, Dakar: 1990, Appendix, for the text of the Yamoussoukro Declaration.

African experts at a meeting held in Mauritius in September 1994, resulted in a series of solutions for achieving the implementation of the Yamoussoukro Declaration, recommended by the experts.[31] The overall recommendation of the experts was to incorporate the Declaration as an integral part of national air transport policy in each African State.[32] The incorporation was to be provisionally effective immediately after the Mauritius meeting.

Later in April 1997, the Banjul Accord for an Accelerated Implementation of the Yamoussoukro Declaration, adopted by Ghana, Sierra Leone and the Gambia, Cape Verde, Guinea Bissau and Nigeria, recognized the region representing these States as a single geographical commercial air transport operations zone for the purpose of implementing the Yamoussoukro Declaration through a joint Secretariat established for the purposes of aeronautical co-operation within these States offering specialized services in air traffic services; safety oversight; market access; engineering and maintenance; communications; aeronautical information services and meteorology.

Pursuant to the Yamoussoukro Declaration in 1988 some progress was made in the aeropolitical scene in Africa. As a result, there is increasing co-operation among airlines of the Southern African Development Community (SADC) countries. These airlines and their countries concerned have already discussed possibilities of operating air transport services by SADC—country airlines under a common logo. The establishing of a Southern Africa Regional Air Transport Authority (SARATA) through a Protocol, was one of the initiatives that was designed to formulate and co-ordinate air transport policies within the SADC region and cope with emergent changes in the air transport industry in the region.

At the global level, the United Nations has made energetic efforts at enhancing development in African States. General Assembly Resolution A/RES/57/7,[33] whilst referring to earlier Resolutions adopted by the United Nations General Assembly,[34] in particular A/RES/56/511 of 15 August 2002 by which the United Nations addressed preparatory issues regarding a high level plenary meeting of the General Assembly to consider ways and means of supporting the New Partnership for Africa's Development (NEPAD), welcomed the NEPAD as a programme of the African Union that embodies the vision and commitment of all African Governments and peoples for peace and development. The Resolution reaffirmed the commitment of the United Nations Declaration on the New Partnerships for Africa's Development,[35] and urged the international community and the United

---

[31] See, United Nations Economic Commission for Africa, TRANS/EXP/94-07 Annex 1.

[32] *Id.* para 2.1.

[33] A/RES/57/7 dated 20 November 2002, Final Review and Appraisal of the United Nations New Agenda for the Development of Africa in the 1990s and Support for the NEPAD.

[34] A/RES/46/151 (dated 18 December 1991); A/RES/48/219 (dated 23 December 1993); A/RES/49/142 (dated 23 December 1994); A/RES/51/32 (dated 6 December 1996); and A/RES/51/216 (dated 21 December 2000).

[35] Adopted at the High Level Plenary Meeting of the General Assembly, held on 16 September 2002.

Nations system to organize support for African countries in accordance with the principles, objectives and priorities of NEPAD, with particular emphasis on the possible contributions in this regard by the private sector and civil society. Arguably, the most significant and useful provision in the Resolution is contained in Clause 32 which requests the multilateral financial institutions to ensure that their support for Africa is compatible with the New Partnership. The United Nations, by Resolution 57/7, also resolved and decided to include a single, comprehensive item on the development of Africa, entitled "New Partnership for Africa's Development: Progress in Implementation and International Support" in the annual agenda of the General Assembly, beginning at its 58th session, while encouraging the efforts being made toward clustering the items related to Africa's development.

In January 2003, at its 57th session, the United Nations General Assembly adopted Resolution A/57/48[36] which addressed issues pertaining to cooperation between the United Nations and the Organization of African Unity[37] including decisions and declarations adopted by the Assembly of the African Union at its first Ordinary Session (Durban, 9–10 July 2002) and earlier United Nations Resolution concerning NEPAD as discussed above. Resolution A/57/48, *inter alia*, requests the United Nations to extend full cooperation and support, as appropriate, to the African Union in its Memorandum of Understanding signed between the Parties on Security, Stability, Development and Cooperation in Africa and requests the agencies of the United Nations system working in Africa to include in their programmes at the national, sub-regional and regional levels activities to support African countries in their efforts to enhance regional economic cooperation in Africa.[38]

African aviation is clearly at the crossroads of progress, where years of enthusiastic endeavour toward consistent cooperation between 53 States are now seemingly giving rise to hope that a liberalized aviation environment will emerge in the African Continent. This is by no means an easy task, given the pluralism and diversity of interests that naturally accompany such a large land mass and multitude of countries. The key word towards success might well be "consistency". Consistency with each others' regulations and uniformity in terms of competitive approach, coupled with the use of geographic, human and other resources would well be the formula for future African aviation.

According to the African Airlines Association, air transport creates directly and indirectly about 470,000 jobs across various sectors and generates revenue of about US$1.7 billion in Africa (2008). African's share of the global aviation market

---

[36] A/RES/57/48 (dated 20 January 2003) Cooperation Between the United Nations and the African Union.

[37] The Organization of African Unity was established in 1963 in Addis Ababa with a membership of 32 African States. It has now 53 independent States. The objectives of the Organization are, *inter alia*, to promote unity and solidarity in African States; to coordinate and intensify their cooperation and efforts to achieve a better life for the African people; to defend the sovereignty and territorial integrity of African States.

[38] A/RES/57/48 supra note 34, Clause 12.

stands at about 5 %. Nevertheless, the African aviation industry, during the recent period of world economic crisis, enjoyed a significant growth rate (5–7 %) compared to other regions of the world. Most significant to this paper is the volume of intra-African traffic, roughly 30 % of the total African air transport traffic, 9.8 million passengers in 2008 and projected to surpass 50 million by 2028.

Given this scenario, ICAO has to concentrate on unity and integration in its efforts in Africa. It has to come to an arrangement (or even an agreement) with the African Civil Aviation Commission, which is the implementing organ of the Yamoussoukro Decision,[39] on a concrete plan to help African States move towards liberalization within their region—which is the fundamental aim of the decision.

References

Abeyratne RIR (2004) The decision in the Ryanair Case – the low cost carrier phenomenon. Eur Transport Law XXXIX(5):585–601
Doganis R (2001) The airline business in the 21st century. Routledge, London
Garvey J (2001) Administrator FAA, The new world of aviation. National Press Club, Washington, DC
Gordenker L (1991) International organization in the new world order. Fletcher Forum World Aff 15:71
Jenks CW (1945) Some constitutional problems of international organizations. Br Year Book Int Law 22:11
Porter ME (1998) On competition. Harvard Business Review Book Series, Boston
Supiot A (2010) A legal perspective on the economic crisis of 2008. Int Labour Rev 149 (2):151–162
White P (1999) The role of the UN specialized agencies in complex emergencies: a case study of FAO. Third World Q 1(20):223–238

---

[39] The Yamoussoukro Decision, which provides for the implementation of the Yamoussoukro Declaration, deals with the liberalization of air transport market access in Africa. Its main objectives are to: (a) facilitate inter-African connectivity; and (b) develop an inter-African network. Its provisions include removal of obstacles, such as restrictions on traffic rights including fifth freedom rights, restrictions and limitations on capacity and frequency between city-pairs, and lack of multiple designation possibilities for competing airlines. Compared to the standard bilateral Air Services Agreement (ASA), the YD gives eligible airlines of all African States fair and equal opportunities to compete based on a common set of harmonized rules and eligibility criteria.

Chapter 2
Competition in Air Transport

Yet another perceived anomaly is that the Chicago Convention has explicitly, in its Article 44, identified three basic aims and objectives for ICAO in the economic field:

(a) Meet the needs of the peoples of the world for safe, regular, efficient and economical air transport;
(b) Prevent economic waste caused by unreasonable competition;
(c) Insure that the rights of contracting States are fully respected and that every contracting State has a fair opportunity to operate international airlines.

All this has to be done under the rubric of ICAO developing principles and techniques of international air transport. Simply put, ICAO has to develop both principles and techniques to prevent unreasonable competition through obtaining for States a fair and equal opportunity to operate international airlines. In other words, ICAO has to ensure reasonable competition between airlines by ensuring that its member States have the ability to give their airlines a fair and equal opportunity.

This objective has been given some direction by the aviation community. During the conclusion of the last session of the Symposium that preceded ATConf/6 (*The Last Word: Strategies and Tools for Sustainability*), the moderator expressed the collective view of the panel as follows:

> ICAO must indeed develop the smart, efficient, effective and harmonized global framework and value-added ground rules to ensure that all the stakeholders of air transport enjoy optimum benefits. Effectiveness would require ICAO and the Sixth ICAO Worldwide Air Transport Conference (ATConf/6) to work towards developing an Annex to the Chicago Convention on the sustainability of air transport.[1]

The possibility of developing an Annex to the Chicago convention is not new. In 1998, when preparing the agenda for the Conference on the Economics of Airports

---

[1] See, The Way Forward: Action Plan for The Implementation of an Improved Regulatory Framework, *ATConf/6-WP/24*, 19/12/12 at 2.

R. Abeyratne, *Regulation of Air Transport*, DOI 10.1007/978-3-319-01041-0_2,
© Springer International Publishing Switzerland 2014

and Air Navigation Services (CEANS), the Air Transport Committee considered the possibility of incorporating the policies and charging principles contained in *ICAO's Policies on Charges for Airports and Air Navigation* Services (Doc 9082) in an Annex. However, since none of the comments received from States in preparation for CEANS addressed this issue, this approach was not pursued. Four key charging principles (i.e. non-discrimination, cost-relatedness, transparency and consultation with users) are already included in national legislation or regulation in a number of States (e.g. The European Union Directive on Airport Charges) and in many air services agreements (e.g. the United States–European Union bilateral).

When discussing the implementation of ICAO policies and guidance in the field of air transport, the Air Transport Regulatory Panel (ATRP/11) recommended that ICAO should conduct a survey, directed at States and concerned organizations, on the relevance and the use of existing ICAO policies and guidance; as well, it was recommended that ICAO should provide States with further information on the proposed coverage and content of a possible new Annex to the Chicago Convention so as to allow States to properly consider the proposal.

A survey conducted by ICAO revealed that, although there was broad support by responding States for the development of an Annex, including coverage of a range of regulatory issues, it was not clear that States have yet reached a consensus on the core substance of such an Annex which is expected to secure the highest practicable degree of harmonization in regulations, standards, procedures, and organization and to help increase awareness and implementation of the existing ICAO policies and guidance related to air transport matters. To pursue the concept of an Annex, it will be necessary first to determine whether a consensus can be developed on the nature and the content of standards and recommended practices governing aspects of the economic regulatory framework for international air transport. A noteworthy point in this regard is that such an Annex would contain standards which will impose some obligation on the part of States to adhere to strict global principles on economic aspects of air transport.

At ATConf/6 a discussion ensued as to whether a new Annex to the Chicago Convention should be introduced as a way forward toward implementing a regulatory framework on economic issues of air transport. The shortcoming of this discussion was that participating States were asked to consider three aspects and their responses were sought to the following:

(a) Whether ICAO should consider the definition of a modular strategy for encouraging the harmonization and modernization of a global air transport regulatory framework;
(b) Whether ICAO should develop a cost benefit analysis for consideration by States regarding the added value of an Annex to achieve this goal; and
(c) Whether ICAO should provide States with further information on the content and coverage of a possible Annex for wider implementation of ICAO policies.

It is submitted that this effort would have proved more efficient if ICAO complied with (c) before engaging in this discussion. The way things stand, it will take another several years for ICAO to come up with "further information" on the content and coverage of a new Annex.

In support of (c) above, Chile, in it submission to ATConf/6 had the valid observation that that a significant influencing factor that perpetuate restrictions on liberalization is that air traffic statistics do not reflect the size of the markets. The statistics collected and processed by ICAO and almost all of the States only record the number of passengers that travel between the origin and destination of a given flight but this number does not always coincide with the origin and the final destination of the passengers. For many international routes, the majority of the passengers on a flight are not travelling to the flight destination but are travelling on further with connecting flights.[2]

Chile said further that, as a result, during bilateral negotiations, States that maintain restrictive policies often call on their counterparts to prove that there has been an increase in traffic that would justify new flights. That is, they insist that their counterparts prove that the size of the air transport market has increased between the two countries. If this increase has mainly come about as a result of connecting rather than direct flights then there is no way of proving this, and thus, their counterparts are unable to obtain an increase in the number of flights.

Therefore, Chile argued that the statistical systems need to be improved to truly reflect the size of markets, which is what is really being negotiated during bilateral negotiations, although one rarely knows the scale of what is being negotiated. However, airlines do have this information. It was proposed that ICAO should revise the criteria that apply to the collection and processing of air traffic.

It was also suggested that ICAO conduct studies on these costs and benefits so that States could see them before bilateral negotiations take place on traffic rights. It is also proposed that ICAO recommend that States conduct these studies themselves on the basis of the guidelines provided to them by ICAO.

2.1 The Safeguards Anomaly

The basic difficulty ICAO faces is in its own misunderstanding of its purpose. If a fundamental task of the Organization, as prescribed by the Chicago Convention, is to ensure that unreasonable competition is obviated, then, prior to adopting an Annex, which in the first place may not be agreed to by the Council, ICAO should recruit experts on competition and task them with producing a comprehensive document on the pros and cons of airlines aligning themselves to market economics on accepted global principles. Such a study could be placed before the Council of ICAO for in-depth consideration as to the added value of having globally accepted principles on liberalization.

---

[2] Proposals for Market Access Liberalization, *ATConf/6-WP/28*.

The ICAO Council has a difficult task. On the one hand, it has to meet the needs of the peoples of the world for safe, regular, efficient and economical air transport. On the other hand, and by the same provision in the Chicago Convention it has to insure that the rights of contracting States are fully respected and that every contracting State has a fair opportunity to operate international airlines. Let us take an example. If a foreign airline offers more connectivity to the world from city A than the national carrier, would it not meet the needs of the people better and more effective than the national carrier which does not have the capacity to offer such connectivity? In this case, should ICAO ensure that the latter has a "fair and equal opportunity to operate air services" from city A, thus stultifying the foreign carrier's services from that city?

ICAO has been continuously adopting a confusing stance of insisting on "safeguard measures" for air transport liberalization. This is an oxymoron by itself as, if there has to be liberalization, why have safeguards that would prove counterproductive to such liberalization? A repetition of this ambivalent approach was once again seen in ATConf/6. In a paper put forward by the ICAO Secretariat at the Conference ICAO claimed that safeguards needed in the air transport liberalization process, covering four specific aspects: the sustained and effective participation of all States in international air transport, assurance of service and State subsidies, essential air services and the issue of unilateral action.[3]

ICAO also claimed that the continuing trend of the past decade towards liberalization has fostered an environment in which States have an increased opportunity for participation in international air transport by allowing more national airlines to operate services, using the services of foreign airlines, or by embarking on joint-venture operations with foreign or same community airlines. This is called creative destruction. Although there are cases where national air carriers have disappeared, the gap is often filled by other airlines. States have adapted to the concept of liberalization and, while the right of participation is no longer a major concern, ICAO guidance on safeguard measures continues to be valid.[4]

Another fallacious argument, perhaps blindly pandering to the nationalistic claims of some States, was that:

> States may need safeguards to ensure continuity of air services to/from their territories. Such a desire is particularly strong for countries experiencing social or economic crisis or where there is an absence of a national air carrier to perform the services, or withdrawal of services, provided by foreign air carriers. Various measures are taken by States to address these concerns including the provision of State aid/subsidies, assistance to national air carriers, particularly for air services of a public service nature, and the conclusion of service assurance arrangements with concerned air operators or parties. In some cases, this support is essential, notably for developing countries and for islands and landlocked countries.[5]

---

[3] Safeguard Measures for Air Transport Liberalization, *ATConf/6-WP/3*, 7/12/12 at 1.

[4] *Id.* 2.

[5] *Ibid.*

This argument sacrifices the day to day activities and functions of a liberalized air transport process by prioritizing the "flag carrier" concept over the needs of the people for true and unfettered connectivity. Besides, nothing in the Chicago Convention supports this neo-colonialist pretension. If a State is in need or in trouble, the international community steps in. An example is the aftermath of the earthquake in Haiti in 2010. The earthquake devastated the capital of Haiti and much of its environs in January, causing massive fatalities and damage to property. The earthquake, crippled the Haitian government and infrastructure, rendering government authorities weak in the running of the country. From an aeronautical perspective, this brought to bear issues of sovereignty within the parameters of relief flights and humanitarian law. Another devastation which was unique to the earthquake was that although the only runway at the airport was undamaged, the rest of the aviation infrastructure lay in a shambles. The flow of the numerous relief flights that came into Haiti after the fact were managed with caution and diligence. Several nations that pitched in with relief in the early aftermath of the disaster did not ponder the legalities concerned with entering the airspace over Haiti or landing therein.

The operation of relief flights, either by States or such bodies as the United Nations, to alleviate human suffering in times of war, natural or manmade catastrophe, is yet another area in which the role of civil aviation is brought to bear in securing peace, security, and connectivity with the rest of the world. There is a specific provision in Annex 9 to the Chicago Convention for provision by State of relief flights. Contracting States are required, by Standard 8.8 of Chapter 8 of the Annex, to facilitate the entry into, departure from and transit through their territories of aircraft engaged in relief flights performed by or on behalf of international organizations recognized by the United Nations or by or on behalf of States themselves and to take all possible measures to ensure their safe operation. The relief flights referred to should be undertaken to respond to natural and man-made disasters which seriously endanger human health or the environment. An emergency is acknowledged in the Annex as "a sudden and usually unforeseen event that calls for immediate measures to minimize its adverse consequences". A disaster is described in the Annex as "a serious disruption of the functioning of society, causing wide spread human, material or environmental losses which exceed the ability of the affected society to cope using its own resources[6]".

The United Nations Charter lists the achievement of international cooperation in solving international problems of an economic, social, cultural or humanitarian character, as one of the purposes of the United Nations.[7] The problems that the United Nations is mandated by its Charter to solve should therefore be necessarily of an international nature. Article 2(7) of the Charter expands the scope of this philosophy further when it provides that the United Nations is not authorized to

---

[6] Annex 9, Facilitation, Tenth Edition—April 1997, Chapter 8, C, Standard 8.8, note 1.

[7] *Charter of the United Nations and Statute of the International Court of Justice*, United Nations: New York, Article 1.3.

intervene in matters which are essentially within the domestic jurisdiction of any State, without prejudice to the right of the United Nations to intervene in matters which are within the domestic jurisdiction of any State, and apply enforcement measures where there is an occurrence of acts of aggression, a threat to the peace or breach thereof.[8] Therefore *stricto sensu*, the United Nations cannot intervene in instances where natural disasters such as famine, drought or earthquakes render the citizens of a State homeless, destitute and dying of starvation unless invited by the States concerned. The principle however cannot be too strictly interpreted, as natural disasters may usually lead to breaches of the peace. In such instances the United Nations Security Council may take such actions by air, sea or land as may be necessary to maintain or restore international peace and security.[9] For such instances, Article 43 of the Charter provides:

> All members of the United Nations, in order to contribute to the maintenance of interna-
> tional peace and security, undertake to make available to the Security Council, on its call
> and in accordance with a special agreement or agreements, armed forces assistance and
> facilities, including rights of passage necessary for the purpose of maintaining international
> peace and security.

Another tenuous argument put forward by ICAO at ATConf/6 on "safeguards" in the same working paper was that in the regulation of air transport services, States often take special measures to ensure that remote communities in their territories are properly served by air in order to encourage development. Such services, of a public service nature, are supported by States in various ways. While essential service schemes are mainly adopted in a domestic context, similar schemes may also be extended at the international level, for example, to promote economic or tourist development in a regional context.[10] Connecting remote communities for internal purposes in the provision of medical and other essential services should not be confused with the meeting needs of the people of the world for international air travel, which is what the Chicago Convention stands for, and what ICAO's aims and objectives are.

Any ICAO study on the role of competition in air transport should take into consideration several factors. In addition to the continued crisis it faces, the airline industry has also been in the throes of a dichotomy. On the one hand, while it has been prolifically international in terms of operations, the airline industry has been unobtrusively national with regard to matters of ownership and control of airlines and interests relating to market access. The latter, brought to bear by regulatory inhibition, which does not admit of an airline freely accessing markets by flying at will to anywhere in the world, and prevailing restrictions as to who owns and controls an airline which bears the nationality of a state, has been increasingly viewed as overtly restrictive in an expanding air transport market. This has led to a gradual liberalization of market access as well as ownership and control in many

---

[8] *Id.* Chapter VII Articles 39, 41 and 42.

[9] *Id.* Article 42.

[10] ATConf/6-WP/3 *supra* note 38 in Chap. 1 at 3.

parts of the world, opening the door to increasing competition between carriers required to cater to a rapidly expanding demand for air transport services. Of course, trends of liberalization and competition have necessitated open skies and common aviation areas to be identified; liberalized air services agreements to be entered into; and legislation enacted, while dextrous commercial tools have been introduced to implement new thinking and circumvent antiquated and counter-productive commercial practices.

At the time of writing, the airline industry was heavily oligopolistic, where three large airline alliances i.e. SkyTeam (made up of Delta, which had merged with Northwest; the merged Air France and KLM and Alitalia), Oneworld (British Airways and American) and Star Alliance (United, which had merged with Continental; Lufthansa and Air Canada). Together, these three major alliances controlled or provided 85 % of transatlantic traffic and fares and 50 % of global fares. Allan Mendelsohn, a former Deputy Assistant Secretary of State of the US State Department had this to say:

> ...It does not seem that the current system is all that different from IATA rate making machinery of years ago, when a select group of international airlines participated and set rates within IATA's then rate-making conferences.[11]

As we are headed towards the next decades, this situation brings to bear the compelling need for corporate social responsibility that would ensure fair and equal opportunity for airlines to compete in providing air transport without being controlled by external factors dictated to by oligopolies. The necessity for corporate foresight stems from the continuing and rapid development of science and technology which are the drivers of social and economic change. Using these two knowledge based and fact intensive fields, airports would be able to obtain a clear picture of challenges and opportunities confronting them. Airlines are a complex, big business and their business environment is highly dynamic. Therefore they need proactive measures to respond to the uncertainties of their business as well as a long term orientation to remain stable amidst imponderables. Airlines need think tanks to mesh their technology trends and market trends to meet a growing demand for air travel. Foremost in this process is a far reaching and forward looking communications strategy as well as a good team of scientific and economic forecasters.

The first step to corporate foresight is to know what the future is going to be like by adopting a foresight-awareness culture. If, as Airport Council International (ACI)[12] Director Angela Gittens said at the 20th World Annual General Meeting of ACI in Bermuda in October 2010, airports should transition from the public utility model to the entrepreneurial business model, the key would be customer

---

[11] Mendelsohn (2012), p. 333.

[12] Airports Council International is the only global trade representative of the world's airports. Established in 1991, ACI represents airports interests with Governments and international organizations such as ICAO, develops standards, policies and recommended practices for airports, and provides information and training opportunities to raise standards around the world.

service excellence. Research and innovation strategies should necessarily be developed through foresight activities. This analogy applies equally to airlines. "Foresight" has been defined as:

> [a] participatory, future intelligence gathering and medium-to-long-term vision-building process that systematically attempts to look into the future of science, the economy and society in order to support present-day decision-making and to mobilise joint forces to realise them.[13]

When one looks at the airline industry and its performance over the past few years, particularly during 2008 and 2009 when the oil price hike took place, it becomes even clearer that airlines will keep facing the vicissitudes of market economics. It is a platitude to say that there is never an easy time in such a competitive and rapidly changing business as the aviation industry. However, it would not be incorrect to say that 2008 and 2009 have been among the most challenging in many years. In 2009, the global economy faced an overarching and systemic recession, leading to significant changes which heavily impacted the industry. Hard times began in the first half of 2008, caused mostly by rapidly escalating fuel prices which peaked at $150/barrel in July 2008.

They became even worse in the second half of 2008 as a result of the credit crunch and the adverse effects on global financial system which nearly collapsed, plunging the world into an abyss of recession which peaked in 2009, the year that registered the first negative growth of the global economy since the great depression of 1929. Some of the hardest hit areas were the air transport, tourism and services sectors.

There continues to be a disconcerting uncertainty as to the depth and duration of this downturn as well as the prospects and benefits of what appears to be an emerging recovery for the air transport sector. Giovanni Bisignani, the down-to-earth, hard-hitting Director General and CEO of IATA, in his address to the Royal Aeronautical Society, Montreal Branch on 1 December 2009 observed that:

> The aviation industry is in crisis. Since 2001, airlines have lost US$53 billion. That includes losses of US$16.8 billion in 2008, the biggest in our history and US$11 billion this year. We are forecasting a further US$3.8 billion loss next year. Look into the detail behind these numbers and you see enormous shifts on key parameters.[14]

Mr. Bisignani correctly attributed this continuing struggle of the airline industry to the mandatory bilateral system of the negotiation of air traffic rights between State authorities, stating that no other industry has such restrictions. He observed that shipping companies operated without restrictions and almost all other products enjoyed global markets where an entrepreneur may have to pay an import tax or a duty but he did not need his government to conclude a bilateral agreement. He added that, however, those airlines could not sell their product where markets exist and could not merge across borders where it made business sense. The result was a sick industry with too many players.

---

[13] Becker (2002), at 7.

[14] http://www.iata.org/pressroom/speeches/2009-12-01-01.htm.

On an earlier occasion, Mr. Bisignani made similar comments, emphasising that numerous businesses have been buttressed by the lowering of trade barriers, but airlines suffered because they still faced enormous hurdles. Mr. Bisignani pointed out that although the airline industry was among the first to operate globally, it was still waiting for the benefits of globalization. He asked a pointed question: "Do we purchase cars or medications based on the nationality of a company's shareholders?" asking the further question that, if an Egyptian can spend a night at a Singaporean hotel in Hamburg, why couldn't an Australian fly a Brazilian airline from Mexico City to Miami? Mr. Bisignani added that airlines supported 29 million jobs and $2.9 trillion worth of economic activity worldwide and that few industries so vital to the health of the global economy remain so restricted by archaic ownership rules.[15]

Mr. Bisignani's successor, Tony Tyler had this to say at the Annual General Meeting of IATA held in Beijing in June 2012, As reported in *Air Transport News*:

> [t]he major benefits of aviation such as connectivity and economic welfares are evident in tremendous contributions of aviation to the global economy such as the provision of 57 million jobs worldwide. However, the state of the industry has been characterised as fragile, with 631 billion dollars revenue but only 3 billion dollar profit for 2012. with high oil prices, political risks around the world and the crisis in the Eurozone a large number of airlines is struggling to keep revenues ahead of costs. Governments and airlines should work as strong partners in order for modern economies to grow prosper and create jobs through being exposed to global opportunities.

Ironically, one could place these facts against the backdrop of pronouncements of high level policy makers who are increasingly reaching the conclusion that aviation is making a substantial contribution to the global economy. *Oxford Economics*, an economic forecasting agency, in a recent report recognizes the wide range of benefits that air transport brings to economies and societies globally.

This report, issued in June 2009,[16] also suggests that the world's future prosperity may depend on a growing and thriving aviation industry, which currently supports nearly 8 % of the world's economy, and questions the environmental benefits and social impacts of limiting that growth. One wonders, then, as to why an indispensable economic tool such as the air transport industry which contributes so substantially to the global economy,[17] is shackled by restrictions which other

---

[15] Bisignani (2006).

[16] Oxford Economics (2009).

[17] ICAO has estimated that the direct contribution of civil aviation, in terms of the consolidated output of air carriers, other commercial operators and their affiliates, was US$370 billion. Direct employment on site at airports and by air navigation services providers generated 1.9 million jobs while production by aerospace and other manufacturing industries employed another 1.8 million people. Overall, the aviation industry directly employed no less than six million persons in 1998. These direct economic activities have multiplier effects upon industries providing either aviation-specific and other inputs or consumer products. In simple terms, every US$100 of output produced and every 100 jobs created by air transport trigger additional demand of US$325 and in turn 610 jobs in other industries. The total economic contribution of air transport, consisting of the direct economic activities and the multiplier effects, is estimated at US$1,360 billion output and

similar businesses are not subject to. It is hoped that the discussion below would shed some light on the issue.

With all this, why is ICAO still speaking of safeguards?

2.2 Competition in Air Transport

It is forecast that the world economy will remain moderately stable and healthy in the near future, despite a slowdown in economic growth. In the short term, inflation may hold steady and inflation rates will probably decrease gradually. The continuing upward trend in fuel prices is likely to increase airline fixed costs and aviation will increasingly be defined in trade terms. Aviation will also be a strong candidate for trade liberalization with a firm focus on services. A compelling factor in this overall picture will be increasing pressure on governments to facilitate transnational ownership of airlines. The other key issue will be aviation and the environment in the global scenario of air transport.

All the above indicators incontrovertibly point to one central driver of future air transport—competition. The issue of competition will ensure the increasing influence of global alliances and partnerships between carriers as a key element in industry strategic development where groups of airlines will provide direction and focus. Airline management, geared towards competition, will be called upon to improve coordination, and provide integration and stability to the air transport industry, resulting in the inevitable corollary of cost reduction.

The outsourcing of non-core activities will continue among airlines, encouraging fledgling carriers to emerge in a liberalized market. Larger airlines will seek franchising and code sharing agreements with other airlines to the farthest extent possible, and will not disregard the importance of creating low cost subsidiaries when possible, while also looking to consolidate their services with other carriers. In the process, existing distinctions between scheduled and non-scheduled (charter) carriers will be minimalized. In terms of service distribution, airlines will invest in e-commerce, while at the same time concentrating as much as possible on selling their services directly on line.

As for regulation, there is a high probability of increasing governmental regulators on safety, security and environment, with heavy focus on the importance of slot allocation for trading services. The demand for aircraft will surge ahead, in keeping with the burgeoning demand for capacity and the compelling need to retire old aircraft for environmental purposes. Very likely, the needed aircraft and engine capacity will be financed by leasing. Organizations such as EU, OECD and WTO,

---

27.7 million jobs worldwide. See The Economic Contribution of Civil Aviation, ICAO Circular 219: 2004 at 1. Also *The Economic Benefits of Air Transport*, 2000 Edition, ATAG:IATA, at 7. Also, *ICAO Circular 219, Id.*, at 4.

whose memberships comprise States, will be called upon to play a greater role in aviation related matters, while States themselves will focus on regulating heavily on consumer rights, environmental protection, security, safety and competition.

Today commercial competition has transcended the past era, where dominant markets protected their established market shares. Most mega commercial activity was then the purview of governmental control under instrumentalities of State which were mostly cumbersome bureaucracies at best. Perhaps the best analogy is the biggest commercial market the United States which had, until recently, extensively regulated larger commercial activities pertaining to energy, transportation and telecommunications.

A global industry, such as air transport, is one in which commercial entities offering their services view competition as global and build strategies accordingly. Therefore, it follows that competition involves a coordinated world wide pattern of market positions, facilities and investments. Factors to be taken into account are overlap between competitors, geographic location of carriers and defensive investments in particular markets and locations so as not to let competitors gain advantages that can be factored into their overall global posture.[18]

Competition in the air transport industry is a complex process, as there is no consensus among airline economists as to the exact nature of the industry. The demand for air services, particularly in the context of the airline passenger, is a contrived demand emerging from other demands based on activities such as business and leisure. This calls for a certain segmentation in travel where, in business travel, the passenger does not usually pay for the travel himself, whereas in leisure travel it comes out of his own pocket. Therefore, the leisure market calls for a different kind of competition, primarily based on the fare, whereas in business travel, although the fare is important, other considerations, such as facilities on board, may also play a considerable role in competition.[19]

Those supporting the retention of regulation argue that the very nature of air transport, being either naturally monopolistic or interdependently oligopolistic, calls for regulation in order that fares are not arbitrarily raised and remain competitive. Another theory in support of regulation is that some form of control should be exercised over mushroom airlines that may sprout up to exploit a liberalized market, thus disturbing the existing balance of an integrated network. Of course, each route is a separate market in itself and would require separate consideration. Although principles of economics of scale may apply generally to airline competition, where a fact such as larger aircraft being more efficient than smaller aircraft would apply on a general basis, individual assumptions for different markets have caused the two major aircraft manufacturers, *Boeing* and *Airbus Industrie*, to concentrate on manufacturing aircraft with strengths in speed and capacity respectively.

---

[18] See Porter (1980) at pp. 291–298.

[19] OECD (1988) at pp. 20–21.

The main consideration, leading up to efforts by the international aviation community to achieve a deregulated global airline industry, is involved with the question as to whether free market principles can be applied globally to air transport. What needs to be considered is whether we are ready to accept the throwbacks as consequences of free market competition in air transport, particularly in losing national prestige projected by flag carriers. One of the corollaries to industry deregulation is the introduction of free market competition when companies switch from operative performance to competitive performance. Competition therefore emphasizes the need to focus on a company and its performance in relation to its competitors. This principle can be readily apply to various industries that have already been deregulated, such as the motor vehicle industry, chemical industry and information technology industry. The operative question is are these good analogies for application to the air transport industry? Whatever be the answer to this question, if the deregulated domestic air transport industry of the United States were to be considered an analogy, one could say that a deregulated system in the United States, introduced in 1978, has led to a more efficient airline system in the country.[20] Whatever be the case, access to facilities in a competitive market is essential toward attaining fluidity of market forces. In the air transport industry, this can be translated to mean that if free markets do not exist in the supply of complementary facilities, there will be no positive impact of liberalization.[21] The complementary services in the supply of air transport are airport access, computer reservation systems and airport and air regulation services.

The International Chamber of Commerce (ICC), in a policy statement has expressed the view that the efficiency of air transport would be enhanced by creating more open markets and more flexibility with regard to foreign ownership. Given air transport's capability to facilitate economic activity, its liberalization would enable the sectors that make use of it to become non efficient.[22] ICC is in favour of a freer exchange of air services throughout the world and is convinced that it is time to move beyond the existing bilateral system, toward a genuine multilateral liberalization of air transport[23] Of course, liberalization would give way to competition, which in turn would impel airlines to pool their resources in order that they maximize on such assets as code sharing and airport slots. However, alliances do not necessarily mean lack of competition between partners. Airlines within alliances have to do their utmost to gain market access and keep their businesses alive. In order to do this both private enterprises and the States in which these enterprises are entrenched have to be equally competitive.

Any agreement to bring in an aspect of trade within a liberalized framework is generally a pro-active measure, which brings to bear the willingness and ability of

---

[20] Bouw and Hall (1999), pp. 147–152 at p. 147.

[21] Banister et al. (1993), pp. 341–348 at p. 341.

[22] The Need to Greater Liberalization of International Air Transport, ICC Policy Statement, Doc 310/504 Rev 3 at p. 1.

[23] *Id.* at p. 9.

the governments to face trading issues squarely in the eye. However, any agreement for trading benefits would be ineffective without the element of competition, both between enterprises and between States. The essential requisite for success in trading relations is competition, which in turn leads to national prosperity. A free trade agreement is merely the catalyst in the process.

2.3   Ownership and Control of Airlines

A stultifying influence on competition in air transport is the ownership and control restriction on airlines. According to this practice, under their bilateral air services agreements, States have traditionally retained the right to withhold, revoke, or impose conditions on the operating permission of a foreign air carrier that is not "substantially owned and effectively controlled" by the designating State or its nationals. This "nationality clause" criterion has been used, since the 1940s, in the overriding majority of air service agreements, and continues to be included in many newly negotiated bilateral accords.

At ATConf/5 held in 2003 the Conference concluded on the subject of air carrier ownership and control that growing and widespread liberalization, privatization and globalization call for regulatory modernization in respect of conditions for air carrier designation and authorization in order to enable carriers to adapt to the dynamic environment. The Conference noted without reservation that, while there are concerns to be addressed, there could also be benefits in liberalizing air carrier ownership and control provisions, observing at the same time that past experience of liberalization in ownership and control has demonstrated that it can take place without conflicting with the obligations of the parties under the Chicago Convention and without undermining the nature of international air transport.

One of the positive considerations that impelled the Conference in its progress on ownership and control issues was that there was widespread support by States for liberalization, in some form, of provisions governing air carrier designation and authorization. In this regard, it was recognized that approaches vary widely from substantial broadening of provisions beyond national ownership and control in the near term, through gradual reduction of specified proportions of national ownership, to limited change for the time being regarding certain types of operations (for example non-scheduled or cargo), application within certain geographic regions, or simply case-by-case consideration. As a result of such varied approaches, there is a consequential need for flexibility in associated regulatory arrangements to enable all States to follow the approach of their own choice at their own pace while accommodating the approaches chosen by others.

The Conference concluded that, whatever the form and pace of liberalization, conditions for air carrier designation and authorization should ensure that safety and security remain paramount, and that clear lines of responsibility and accountability for safety and security are established for the parties involved in liberalized arrangements. Economic and social consequences of liberalization were also considered by the Conference at some length. The Conference recommended that in liberalizing the conditions for air carrier designation and authorization, States should ensure that the economic and social impact, including the concerns of labour, are properly addressed, and that other potential risks associated with foreign investments (such as flight of capital, uncertainty for assurance of service) are fully taken into account.

The rationale for the nationality clause is that it provides a convenient link between the carrier and the designating State by which parties to the agreement can: (a) implement a "balance of benefits" policy for the airlines involved; (b) prevent a non-party State through its carrier from gaining, indirectly, an unreciprocated ("free rider") benefit; and (c) identify the country that is responsible for safety and security oversight. National defence considerations are also a factor in some cases. The nationality clause made obvious sense in the days when most airlines were State-owned.

Over the past two decades, liberalization, privatization and globalization have significantly changed the airline industry. Transnational investments in air carriers have occurred against a backdrop of widespread multinational ownership in other service industries. The original bases for use of the nationality clause have been seen as increasingly at odds with the changed global business environment in which the industry must operate. Past analysis in ICAO and considerable State practice in recent years have confirmed that safety and security can be safeguarded without reliance upon the traditional nationality clause.

To facilitate liberalization, ICAO has addressed this issue extensively and developed guidance for States. Although there has been some encouraging progress in State practice in terms of relaxing the application of the rules or accepting airlines with foreign ownership, legal restrictions on ownership and control of airlines in States' laws and bilateral agreements have largely not changed. Ten years after the 2003 Fifth Worldwide Air Transport Conference (ATConf/5), continuation of such legal restrictions is often characterized as abnormal and harmful. The need to enable air carriers to adapt to the dynamic global environment and to enable States to participate more effectively in international air transport calls for a change in approach.

Facing heightened competition on the one hand and the constraints of the regulatory system on the other, airlines have found various ways by which to expand networks across national borders, including alliances, mergers and acquisitions, many of which involved foreign investment or equity exchange. Recent years have seen an extraordinary acceleration of this trend. In Europe, Latin America, and Asia, airlines with foreign or transnational ownership and control, as well as "families" of related carriers in different countries but under a single management, are now a common and growing phenomenon. Although there have been some accommodation in the legal framework for these developments, the mergers and acquisitions yielding these new airline enterprises have, in most cases, occurred without full legal protection from the potential inherent in nationality clauses to block or condition air services by the new entities. In many cases, airline managements have felt compelled to adopt complex "holding company" structures or create different classes of stock ownership in order to maintain a legal defence against the invocation of nationality clauses.

Along with industry changes, some States have adopted more flexible treatment of the airline ownership and control requirement. As demonstrated by State practice over the past decade, an increasing number of countries are now more willing to accept air services operated by airlines that have substantial foreign ownership or are multi- or trans-national in character. The case that 114 States have now accepted the "Community carrier" principle of the European Union (EU) through "horizontal agreements" is one, but not the only, prominent demonstration of this important change. Other examples include the use of alternative criteria for airline designation and authorization in bilateral agreements and acceptance of cross-border mergers and acquisitions although mostly within a regional grouping of States.

In 2009, seven States, through the platform of the International Air Transport Association (IATA) "Agenda for Freedom" initiative, took coordinated action by signing a "Statement of Policy Principles" in which the States undertook, as a political commitment, to liberalize key aspects of international air transport regulatory practice, including airline ownership and control by waiving the nationality clause "on the basis of reciprocity". This statement, endorsed by the European Commission, has since been signed by four additional States, and remains open for endorsement by any interested State. At the 37th Session of the ICAO Assembly, IATA expressed the view that ICAO should now resolve to move this initiative forward.

Drawing from an ICAO recommendation, the United States (U.S.) took the initiative to explore with interested aviation stakeholders the possible formulation of a legally binding instrument for liberalizing ownership and control. At the 37th Session of the ICAO Assembly, the U.S. proposed that ICAO consider the development of a multilateral convention to facilitate airline access to international capital, which the ICAO Council later agreed to include in its work programme in relation to the preparation of ATConf/6.

ICAO has considerable policy guidance on the subject, consisting of a number of alternative criteria and model clauses for optional use by States which are available in air service agreements, Assembly resolutions, conclusions and recommendations of the air transport conferences (contained in the *Policy and Guidance Material on the Economic Regulation of International Air Transport*.[24] The guidance provides States with useful options to liberalize ownership and control rules, whether on a unilateral, bilateral or multilateral basis. The coordinated action taken by several like-minded States cited in this section above is an example of one option available to States. ICAO also implemented some of the 2005ATConf/5 recommendations to facilitate liberalization in this area, such as collecting and publishing States' relevant policies, positions and practices. Despite some progress and the continued validity of the ICAO policy guidance, the lack of widespread use of such guidance suggests that other options may need to be explored.

---

[24] Doc 9587.

For a long and sustained period of time since the formal regulating of civil aviation, many countries have owned and controlled their national carriers, partly for national prestige and symbolism, and partly because of a traditional requirement in the standard bilateral air services agreement, that a designated carrier should be substantially owned and effectively controlled by nationals of a country which designated that carrier to operate air services under bilateral agreements entered into by that country. This requirement, although tolerable in the first decades of commercial aviation when demand for capacity was manageable, gradually evolved into being an inhibitor in the provision of air transport services. Many States were left with unprofitable State owned airlines that largely required subsidization. The circumscribing nature of an inflexible ownership and control requirement has prompted many States to permit privatization of air carriers, with a reduction in percentage of government held shares. For example, British Airways and Lufthansa have been completely privatized, while Air France, Alitalia, Sabena and Iberia have been partially privatized. Across the Atlantic, the United States deregulated its domestic carriers in 1978 in order to meet capacity demands efficiently and equitably, offering lower prices and improved services to customers.

Although liberalization of air transport is sweeping the globe with its various attractions, offering more capacity and competitive services, it is claimed that the bilateral air services agreement is still preventing pro active measures of airlines to merge with each other and enter into other strategic alliances, through antiquated requirements of national ownership.[25] The bilateral requirement of substantial ownership and effective control, which is based on the fundamental postulate that a majority ownership provision would effectively preclude foreign ownership from taking major control of a national carrier, has not been easy to enforce or put into practice in all situations. Although a blanket provision might require majority national ownership and control, airlines and States have had to contend in many instances with complex issues of nationality of members of a board of directors, the powers of a board and the powers of directors of such boards.[26] Often States have attempted to circumvent these difficulties by establishing a safeguard to ensure for the government concerned a golden share which accords the owner government a greater voice in the decision making process on issues of importance and signifi-cance to the carrier concerned.[27]

Although airline alliances may offer a way round circumscriptions of the market access constraints that may be presented by bilateral air services agreements, such alliances are not usually effective against the inhibiting qualities of the traditional ownership and control provisions of the typical agreement, particularly in the context of facilitation of cross border investment, which is essentially regulated

---

[25] See Bohmann (2001), at p. 690.

[26] See ownership and Control, Report of the Think Tank—World Aviation Regulatory Monitor, Geneva, 7 September 2000, prepared by Peter van Fenema, IATA, Government and Industry Affairs Department at p. 4.

[27] *Ibid.*

by the bilateral air services agreement. In order to find a practical and legitimate way out of this seemingly impossible situation, ICAO has devised a pro active approach based on making the principal place of business and permanent residence of the carrier the operative criteria for purposes of devolution of control.

Both the United States and member States of the European Union have protected their domestic markets from external operators by preserving these markets for their flag carriers or at least carriers that were owned by the State concerned or nationals of that State. In the European Union, in keeping with Article 4 of Court Regulation 2407/92, national authorities are vested with the power of granting operating licences based on the criteria that the principal place of business of the carrier applying for the licence must be located in the licencing member State; the carrier must be involved in air transportation as its main occupation; the holder of the licence must be under direct or majority ownership nationals of the European Union; and the licence must be effectively controlled by such nationals. Effective control essentially means the power and ability to exercise a decisive influence on an air transport undertaking, including but not limited to the use, enjoyment and alienation of movable and immovable property of that undertaking. One of the reasons, at least from the perspective of the European Union, of retaining the ownership and control criteria within its territory, is to safeguard the interests of the member States of the Union and to preclude carriers of non EU States from capitalizing on a liberalized European Union Market.

United States law too contains explicit requirements pertaining to nationality in terms of management of airlines,[28] in some contrast to Regulation 2407/92 of the EU which does not expressly address issues regarding nationality of management. Arguably, the EU addresses external control by stockholders of a company, and not particularly, as envisaged by the United States law, management control of the administration and running of the air transport enterprise.

Be that as it may, both the United States and the European Union have shown, by their legislation, that the issue of ownership and control still remains to them a critical consideration in the overall picture of liberalization of and competition in air transport.

At ATConf/6, the European Union argued:

> In today's more and more globalised and competitive environment in which air carriers operate, it is worth for ICAO to examine whether the above reasons are still valid and indeed whether such restrictions serve the interest of air carriers or indeed public interests, taking into account the vast potential benefits that liberalization may bring. The policy objectives for restrictions may be better served by other, more flexible means, for example through national licensing requirements that would not deprive the air transport sector of the potential benefits of liberalization[29]

The EU further argued that liberalization would allow increased access by air carriers to capital which will become increasingly important in a more and more

---

[28] 49 U.S.C. 40102 (a) (15) 1994.

[29] National Restrictions on Air Carrier Ownership and Control, *ATConf/6-WP/50* at 2.

competitive environment, and which would in general facilitate the rationalisation of the air transport sector which is still subject to restrictions unknown in other economic sectors. Another point in favour of doing away with nationality restrictions is that free investment in air carriers could support consolidation, mergers and takeovers in the air transport industry, and could enable the development of transnational air carriers that are able to serve the whole air transport network worldwide without being subject to unbalanced restrictions. Besides liberal designation and traffic rights regimes in ASAs (i.e. multiple designation of air carriers and unlimited third and fourth freedom rights at least), the increased possibility of establishing or investing in air carriers by foreign nationals would possibly contribute to further market opening, and thus play a supportive role in the process of liberalising air traffic rights.

Another compelling point considered by ATConf/6 was that the possibility of foreign investment would support the privatisation of nationally owned air carriers in cases where States decide upon privatisation. Where national air carriers can accept foreign capital, their reliance upon state aid and subsidies may be reduced. This would reduce the burden on governments to continue financing nationally owned air carriers and would contribute to improving air carrier management as investment by foreign air carriers may also bring new know-how and best practices. As European experience shows, the liberalisation of air carrier ownership and control may help improve the economic efficiency of the air carrier industry.

In order to face the exponential growth of the air transport industry, it is inevitable that competition and liberalization of the industry should be given serious consideration as a current and future trend in the aviation field. What is needed foremost, to improve international cooperation toward achieving a well meshed and overall competitive policy, is to consider the various possible options available. One of the options to promote competition and facilitate trade in air transport lies indisputably in combatting and eliminating anti competitive practices. State responsibility toward achieving this goal is a key factor. One way of ensuring collective State action in this regard and enforcing the duties of States toward the international community is to establish an international entity charged with ensuring the implementation of a global competition code. States may also enter into understandings or agreements toward combatting restrictive trade practices, either bilaterally or plurilaterally. Along with a plurilateral framework of competitive policy, there also should be a concomitant bilateral structure of individual agreement between States to stringently monitor anti-competitive conduct.

As for liberalization of air transport, there has so far been no indication whatsoever that any State favours total liberalization calculated to open out its domestic market. Strategic alliances between airlines, be they through mergers or other arrangements, will be viewed with caution and objectivity by individual airlines and States so as to preclude the total overrunning of local interests. It is this consideration that would make liberalized ownership and control criteria less attractive to local entrepreneurs who would not encourage foreign ownership to encroach local control airlines have of their own markets.

Finally, although liberalization of air transport cannot be dismissed as a viable prospect for the future, particularly in trading terms, the players concerned must necessarily view air transport in its entirety, as a service composed of critical factors that are inherent in safe and efficient air transport.

At ATConf/6 ICAO made some perceptive observations and constructive suggestions. ICAO advised the Conference that the majority of States have continued to use the traditional nationality clause in bilateral agreements, not only leaving it unchanged in existing agreements but also adopting it as part of new air service accords. In part, this may reflect the hesitancy of States to endorse new language in an area long governed by tradition. In addition, there is evidence that States are reluctant to adopt an approach that does not easily distinguish between the nationalities of third-country investors. For example, the law of the EU permits ownership and control of EU air carriers by nationals of third countries only if such countries and the EU have concluded an appropriate, reciprocal agreement. The U.S. has waived the nationality clause in cases where both the relevant airline and the foreign investors are covered by a U.S. negotiated "open skies" agreement, although the U.S. is generally reluctant to legally bind itself to this policy.

Accordingly, there is merit in exploring more practical solutions which would enable States to deal with the ownership and control requirement with flexibility in light of other policy considerations and without the arduous requirement to amend every air services agreement so as to provide greater certainty to States from the regulatory perspective, and to air carriers making strategic and investment decisions.

While States can continue to use the various means recommended by ATConf/5, contained in *Report of ATConf/5*,[30] to advance this goal, ICAO should play a leading role in exploring alternative solutions to modernize the regulatory regime so as to better adapt it to a globalized and liberalized business environment, thus meeting the needs of airlines to operate as a "normal industry".

In this regard, one action ICAO can take is to explore the development of an international agreement for States to relax ownership and control requirements for airline designation and to facilitate airline access to international capital. A recent survey of States in October 2012 revealed that 77 % of the responding States (47 of 61) were supportive of ICAO taking this initiative. The development of such an agreement could draw on past experiences and build on what has been achieved, such as the agreement reached by States under the IATA Agenda for Freedom Summit initiative.

ICAO observed that one option is for the agreement to take the approach of a "waiver of nationality clause", whereby parties to the agreement, on the basis of reciprocity, commit legally to waive the application of the nationality clause in existing air services agreements in respect of designated airlines and investing nationals of other concerned States. Another option could be for the parties to

---

[30] Doc 9819.

accept and apply a common criterion for airline designation, such as the "principal place of business and effective regulatory control" criteria developed by ICAO. The design of the agreement should reflect a vision to move forward yet be based on reality, taking account of States' willingness to adapt to change on an incremental basis and constraints in respect of changing national laws. The agreement should therefore not impinge on any State national legislation and policy in this regard. In light of the disparity in States' needs and circumstances, the agreement could be for signature by "willing and ready" parties initially and subsequently open for accession by other parties.

In any case, the development of such an agreement would require substantive and incremental work by ICAO, including further study of the options, and consultation with experts and States and regional groupings. The challenge is for the international aviation community to agree that the time has come for more concrete exploration of such an agreement, and for ICAO to work with States and the industry to define the essential elements that would garner the broadest support and contribute most effectively to the sustainable development of international air transport.

The conclusions that one can reach from the above discussion are that:

(a) Since ATConf/5, although diverging views and regulatory approaches remain with regard to air carrier ownership and control, more States are willing to liberalize as evidenced by the granting of substantial ownership in State airlines and acceptance of designations of other States' airlines with majority foreign ownership and effective control, notwithstanding the nationality clause in their bilateral agreements;

(b) Considerable progress has also been made at the regional level, as States in several regions or sub-regions have adopted regional arrangements in liberalizing air carrier ownership and control among respective members, allowing transnational airline mergers to proceed;

(c) Despite some progress in liberalization and the continued validity of ICAO policy guidance, States are hesitant to make globally applicable policy statements or commit to allow all airlines from all States to have unlimited foreign ownership. There is a need to explore other more flexible options that can achieve wider acceptance and allow liberalization to move forward, including in an incremental manner, without affecting national legislations; and

(d) ICAO could play a leadership role in developing policy guidance and in facilitating regulatory evolution, including considering the development of a multilateral agreement to meet the needs of States and industry, as recommended by ATRP/11.

In this context, ICAO's suggestions were:

(a) States should continue to liberalize air carrier ownership and control through various means, including those recommended by ICAO;

(b) ICAO should continue to promote its policy guidance on air carrier ownership and control, and encourage States to use its guidance in regulatory practice.

It should also keep its policy guidance current and responsive to changing situations and requirements of States; and

(c) ICAO should initiate the development of an international agreement for States to liberalize air carrier ownership and control.

ICAO's responsibilities in this context are quite daunting although not unattainable. Since the 5th Conference, at the time of ATConf/6 (March 2013) scheduled passenger traffic had increased by 70 %. This figure was expected to double by 2030.[31] Aviation is a catalyst for growth and by 2013 the world community expected 3 billion passengers to have been transported by air during the year. Oxford Economics, a global forecasting and research consultancy, estimates that aviation's global economic impact (including direct, indirect, induced, and tourism catalytic effects) amounts to roughly US$2.2 trillion. In addition to contributing 6.6 million jobs worldwide, aviation supports 3.5 % of global Gross Domestic Product (GDP). Consequently if aviation were a State, it would rank as the world's 19th largest country in terms of GDP. Based on the UAE's experience, this paper presents some recommendations to promote liberalisation in air transport worldwide and thereby enhance and fully realise the vital contributions of aviation to global economic development and prosperity.

In order to cope with thus increasing trend, ICAO has to start with its existing "guidelines" on economic policy.

2.4   Existing Guidelines

ICAO Member States are expected to comply with ICAO Standards and Recommended Practices (SARPs) where States cannot comply, the State must file a difference before the effective date of compliance with the SARPs. However, ICAO policies in the field of economic regulation are adopted at a global level by ICAO Member States and while these policies are agreed upon by the international aviation community and, as such, imply a moral obligation for State compliance, the policies are not mandatory; application, implementation or compliance are at the States' discretion.

Over the years, the ICAO Assembly recognized the importance of such policies and, in Assembly Resolution A37-20—*Consolidated statement of continuing ICAO policies in the air transport field*, urged States to give due regard to the policies and guidance material developed by ICAO on economic regulation of international air transport. Accordingly, the emphasis of the Secretariat work pursuant to the 1994 and 2003 Air Transport Conferences was on promoting and assisting States in implementing a range of ICAO policy and guidance on economic regulation.

---

[31] Benjamin (2012), at 5.

Existing ICAO policy and guidance material on the regulation of international air transport is contained mainly in two publications: Doc 9587, *Policy and Guidance Material on the Economic Regulation of International Air Transport*, and Doc 9626, *Manual on the Regulation of International Air Transport*.

Doc 9587 is a comprehensive document of all formal policies and guidance material adopted by ICAO concerning the economic aspects of international air transport regulation. In addition to relevant Articles of the *Convention on International Civil Aviation* (Chicago Convention), the International Air Services Transit Agreement, Doc 9587 also contains relevant Assembly and Council resolutions, decisions, as well as conclusions, recommendations and declarations of air transport conferences. It also contains Council endorsed guidance material, such as the ICAO Template Air Services Agreements (TASA), and recommendations adopted by the Air Transport Regulation Panel (ATRP). The present version of Doc 9587 is the third edition, published in 2008.

Doc 9626 both complements and supplements Doc 9587. It contains descriptive and analytical material on existing regulatory processes and structure at the national, bilateral and multilateral levels, as well as regulatory content and key issues. Doc 9626 provides a comprehensive and objective source of information regarding the many facets of international air transport regulation. It provides useful reference material for regulatory practicioners, and for training purposes. The current version is the second edition published in 2004.

Information on the adherence of States to the ICAO policies on taxation is made available to States in the Supplement to Doc 8632. As for ICAO policies on charges, information is made available through a report on the implementation of the recommendations of the 2008 ICAO Conference on the Economics of Airports and Air Navigation Services (CEANS), Doc 9908. Both policies and supplementary information are made available on the ICAO public website. It is to be noted that more than half of the Member States do not provide ICAO with information on adherence to the policies. While the majority of States reporting such information adhere to the policies, some reported partial adherence, and others did not indicate the level of implementation.

With regard to the importance and use of ICAO guidance material by States, the ICAO Secretariat often receives positive feedback from States through the ICAO Regional Offices or at workshops and seminars. Some States have reported the incorporation of ICAO model clauses in bilateral air services agreements. However, the extent of the reporting has not been quantified in a systematic manner. Likewise, some industry organizations also supported and promoted the use of ICAO policies and guidance.

As outlined above, ICAO addressed many of the regulatory issues and produced a wide range of policies and guidance material to ensure the safe and orderly growth of international air transport. As recommended by ATRP/11, a survey of States on the relevance and use of ICAO policies and guidance was conducted in October 2012 and revealed that over 95 % of the responding States (58 out of 61) were aware of the existing policies and guidance material and over 83 % (51 out of 61) considered the documentation relevant. Some States suggested that ICAO should review and update the guidance material to take into account new developments in regulatory practices.

The survey demonstrates that the main problem is not a lack of awareness of ICAO policies and guidance or their relevance but rather a lack of implementation, probably due to their non-binding nature. Such policies were developed to promote common interest for the international aviation community. Adherence to the policies is in the best interest of States, promotes the efficient and orderly development of international air transport, and is contributing effectively to trade and economic development. The lack of implementation indicates a compelling need to find ways by which to give more "teeth" to the ICAO policies, and increase efforts to encourage State implementation. Hence, the access to, and promotion of, the policies is an issue that requires particular attention if the policies are to remain of relevance and value.

It is worth noting that the 2008, CEANS considered a similar agenda item regarding the implementation of ICAO policies on charges and the need for States to be "morally committed" to the policies. The Conference concluded, inter alia, to encourage States to adopt the principles contained in Doc 9082 in national legislation, regulation or policies and to incorporate them into future air services agreements so as to ensure compliance by airports and ANSPs. These conclusions have also been considered as important measures to enhance adherence to, and implementation of, the ICAO policies on taxation (see ATConf/6-WP/10).

With respect to ways to promote the policies and guidance material, regional workshops and seminars are effective tools not only for States to learn about the ICAO policies and guidance, but also for the ICAO Secretariat to obtain feedback on the policies and on State needs. These events also provide an excellent opportunity for States to exchange information on how to best implement the policy guidance and how to overcome implementation difficulties. It should be noted, however, that the reduced number of workshops/seminars may impact the ability of States to implement ICAO policies and guidance material.

In addition to workshops and seminars, the Secretariat has been exploring, with partners in the industry, the development of a series of appropriate training courses to promote the use and implementation of ICAO policies and guidance in the air transport field. In that context, ICAO is currently developing with CAE1 e-learning courses on air transport regulation. Electronic dissemination has become an important means by which to provide information and also an easier means for States and users to report to ICAO on implementation of the policies by States and other users. ICAO could improve its use of these measures as possible means to promote awareness and understanding of the relevant policies and guidance.

It is essential that ICAO policies in the economic field be given more "teeth", as this will support the role ICAO is to play for the sustained economic development of the air transport system. In this context, re-affirming this role is imperative so that ICAO can continue to assist its Member States in developing policies and practices that facilitate the globalization, commercialization and liberalization of air transport.

The leadership role of ICAO in economic regulation of international air transport, and in the development of comprehensive policy guidance to assist States in

the creation of a favourable regulatory environment for the sustainable development of air transport and the benefit of all stakeholders, is indisputable. ICAO
policies and guidance material on the economic aspects of international air transport regulation remain relevant but need to be kept current and responsive to
changing situations and the needs of States and aviation stakeholders. There is a
need for ICAO, in cooperation with the industry, to ensure widespread awareness
and improved implementation of its policies as well as use of its guidance material
on economic regulation.

In view of the above, and as a starting point, States should recognize the
importance and relevance of ICAO policies and guidance and give due regard to
them in regulatory practices and should exert all effort to ensure adherence to
commitments relating to provisions of Assembly Resolutions in the air transport
field. For its part, ICAO should encourage States to incorporate ICAO principles,
policies and guidance in national legislations, rules and regulations, and in air
services agreements and continue to promote its policy guidance on the economic
regulation of international air transport, and encourage States to use ICAO guidance
in their regulatory practice. ICAO should also ensure that policies, guidance and
other material related to economic regulation remain relevant, current, and responsive to changing situations and requirements of States and continue to consider
additional ways and means by which to enhance the status of its policies for the
sustained development of the air transport system.

As a final development and the end result of ICAO's work, the Organization
suggests:

> Another way to overcome the weaknesses of the bilateral system and improve the global
> regulatory process and structure is for ICAO to assume a leadership role in developing a
> multilateral approach to facilitate market access expansion. It may take the form of an
> international agreement under which parties commit to remove restrictions on core
> Freedoms of the Air. The development of such an agreement could benefit from past
> experiences and build on what has been achieved, such as OSAs, regional liberalization
> arrangements, and plurilateral or multilateral agreements. In light of the disparity in States'
> needs and circumstances, such an agreement could be designed for signature by the
> "willing and ready" parties initially, for instance, by States already having OSAs with
> each other, and open for accession by any ready parties subsequently.[32]

Although this a laudable and proactive approach, it offers some confusion with
the concept of the new Annex suggested. If the "multilateral approach" referred to
suggests a multilateral treaty ICAO would have to carefully think through the
position of such a treaty under the Chicago Convention. If the new treaty were to
amend the Chicago Convention it must be approved by a two thirds vote of the
Assembly and then come into force in respect of States which have ratified such
amendment when ratified by the number of contracting States specified by the
ICAO Assembly. The number so specified cannot be less than two thirds of the total
number of contracting States.[33]

---

[32] Expanding Market Access for International Air transport, *ATConf/6-WP/13*, at 3.

[33] Chicago Convention, *supra*, Article 94.

If a new treaty is pursued, international treaty law will apply. A treaty enters into force when the number of ratifications as specified in that treaty is received by the depository. When a treaty enters into force it is in force for only those States who have consented to be bound by it which are called "Parties".[34] However, an expression by a State that it consents to be bound by a particular treaty does not mean that *ipso facto* that treaty enters into force for that State. Either, the treaty must already be in force at that time, or as already mentioned the number of ratifications must be deposited. The Vienna Convention (1969) is more specific when it says that a treaty enters into force in such manner and upon such date as it may provide or as the negotiating States may agree[35]. There are three ways in which a treaty may enter into force. They are: on a date specified in the treaty; on signature only, as agreed by the negotiating States; or on ratification by all or a specified number of States. A treaty may be considered to apply to a State provisionally when the treaty itself so provides; or the negotiating States have in some other manner so agreed. Unless the treaty otherwise provides or the negotiating States have otherwise agreed, the provisional application of a treaty or a part of a treaty with respect to a State will be considered as terminated if that State notifies the other States between which the treaty is being applied provisionally of its intention not to become a party to the treaty.[36]

It must be noted that the Chicago Convention also provides that any contracting State may make arrangements not inconsistent with the provisions of the Convention and that any such arrangement shall be forthwith registered with the Council, which shall make it public as soon as possible.[37] Furthermore it requires that contracting States accept the Convention as abrogating all obligations and understandings between them which are inconsistent with its terms, and that they undertake not to enter into any such obligations and understandings. A contracting State which, before becoming a member of the Organization has undertaken any obligations toward a non-contracting State or a national of a contracting State or of a non-contracting State inconsistent with the terms of this Convention is required to take immediate steps to procure its release from the obligations. If an airline of any contracting State has entered into any such inconsistent obligations, the State of which it is a national is obligated to use its best efforts to secure their termination forthwith and to cause them to be terminated in any event as soon as such action can lawfully be taken after the coming into force of the Convention.[38]

---

[34] Vienna Convention on the Law of Treaties, 1969, Done at Vienna on 23c May 1969, United Nations General Assembly Document *A/CONF.39/27*, 23 May 1969, Article 2 (1) (g). The Convention entered into force on 27 January 1980. UNTS Vol. 1155, p. It should be noted that such States should not be called "signatories" as some refer to them erroneously.

[35] *Id*. Article 24 (1).

[36] *Id*. Article 25.

[37] Article 83 of the Chicago Convention.

[38] Article 82 of the Chicago Convention.

2.5 A Compromise

Both a multilateral treaty on liberalization and a new Annex to the Chicago Convention will have their own problems. A multilateral treaty will be a cumbersome project and may not easily pass through the majority of the 191 ICAO member States. On the other hand, a new annex is an internal affair within the ICAO Council. However, the Council has shown considerable reluctance to elevate the economic regulation air transport to Standards under an Annex.[39] However, if, as New Zealand suggested at ATConf/6 the idea is to develop a "multilateral agreement" it would be a completely different prospect. Such an instrument would provide text for more than two States to enter into an agreement for market access and obviate the thousands of bilateral arrangements now being used which is an inefficient way of exchanging air rights leading to uncertain airline planning and a lack of standardisation of requirements. This inefficient system has done little to strengthen the sustainability of an international airline industry that has a notoriously poor record of financial performance. The system has also supported protectionism and inefficiency at the expense of transport users.

Whilst there has been a global trend towards liberalisation, in considering what form a new global instrument or instruments should take, New Zealand suggested that there were two key requirements. First, to enable greater competition, along with the removal of route and tariffs restrictions, the artificial distinction between third/fourth freedom and fifth/sixth freedom traffic should be dropped. This should be accompanied by the removal of capacity limits to allow greater competition for passenger and cargo traffic. Second, the threat that an airline might be denied operating authorisation on foreign ownership and control grounds should be removed so that the financially risky airline industry has open access to global capital markets and skilled management. The problem with this system would be that States may not feel obligated to use such an instrument, which will be relegated to the filing cabinets of States' civil aviation archives.

The compromise might lie in a formal interpretation of Article 6 of the Chicago Convention which provides that no scheduled international air service may be operated over or into the territory of a contracting State, except with the special permission or other authorization of that State, and in accordance with the terms of such permission or authorization. This provision speaks only of permission to operate over or into the territory of a Contracting State. The Council may wish to

---

[39] It must be noted that the Arab States presented its case for a new Annex at ATConf/6. The States said that although there are various guidance material of ICAO on the economic regulation of air transport, "the international air transport industry urgently needs a new Annex that addresses international air transport regulatory matters and defines its requirements in a binding legal text through Standards and Recommended Practices, while taking into consideration developing countries interests in accordance with the ICAO recommendations as indicated in this paper. It should be noted that the Arab Member States of the Arab Civil Aviation Commission (ACAC) endorse and support this step". See ATConf/6-WP/27 at 2.

suggest an amendment to this provision, to the effect that permission by a grantor State will not be refused if such operation facilitates and improves connectivity and meets the needs of the people for air transport. The onus of proving this would be on the applicant airline through its State of nationality.

Under this interpretation of the revised or amended Article 6, if justification is warranted, a state could grant the air carrier of another State the right to provide scheduled and non-scheduled passenger and cargo or all-cargo services between the territories of the Parties (3rd and 4th Freedoms), between the territory of the other Parties and any third State either directly (5th Freedom) or transiting through its home territory (6th Freedom), with no requirement for the cargo services to include any point in the territory of the State designating the airline (7th freedom); without any restrictions as regards routes, frequency and flight material, which may be owned, leased or chartered.

The United Arab Emirates submitted the following statement at ATConf/6 in support of this premise:

> For decades, the UAE has recognised the above-mentioned benefits and therefore made liberalised aviation a strategic policy imperative. Thanks to openness to competition aviation has become a core sector of the UAE's economy, driving development, diversification and aviation-related activities which are contributors to the UAE's non-oil GDP. Aviation links the UAE to the world, providing connectivity that is vital for a young country that was only established on 2 December 1971. In this context, every policy objective must recognise the strategic contribution of the open and liberal sector in fostering the country's economic goals. In other words, aviation is a catalytic element of the economic supply chain and a vehicle for achieving economic and social development. The UAE aviation sector is growing rapidly because of the success achieved by the domestic and foreign airlines in stimulating new and existing international and intercontinental flows to and from the UAE that, in turn, stimulate the development of local businesses, industries and trade. Consistently, the UAE has therefore endorsed a liberal aviation policy, both in fostering competition within itself (with numerous airlines and airports) and in its firm commitment to pro-competitive Open Skies air services agreements. To date, the UAE has air services arrangements with 147 countries, of which 113 are "Open Skies" or fully liberal agreements[40]

Under such a policy, States involved may have the right to designate as many airlines as they wish to operate air transport services under such an agreement and to revoke or change such designations. These designations should be required to be sent in writing to the corresponding Party through diplomatic channels, as well as to the Depositary. The aviation authorities of one Party may require that an airline designated by another Party demonstrate that it is qualified to meet the conditions established by the standard laws and regulations reasonably applicable by these authorities for air transport operations. Each Party should have the right to withhold or revoke the designation or to impose conditions upon a designated airline which it deems necessary for exercising the rights specified in Article 1 of this Agreement, if the airline is not incorporated and does not have its main office in the territory of the designating Party or if this Party does not have the regulatory control.

---

[40] Views on Advancing ICAO's Work on Air Transport Liberalization, *ATConf/6-WP/33*, at 2.

With regard to open competition, which is the most critical feature of liberalization, each designated airline should be free to determine the level of transport capacity which they wish to provide. Neither State Party should be entitled to unilaterally restrict the traffic volume, frequency or scheduling of the service or the type or types of aircraft operated by the airlines designated by the other Party, except where necessary for customs, technical, operational or environmental reasons, in accordance with the standard conditions in line with Article 15 of the Chicago Convention and always on a non-discriminatory basis.

Globalization of competition is one of the key messages of the modern age, which brings with it the important message for governments to rethink their strategies with regard to the air transport strategy as an integral part of their national competitiveness. Any agreement to bring in an aspect of trade within a liberalized framework is generally a pro-active measure, which brings to bear the willingness and ability of the governments to face trading issues squarely in the eye. However, any agreement for trading benefits would be ineffective without the element of competition. The essential requisite for success in trading relations is competition, which in turn leads to national prosperity. A free trade agreement is merely the catalyst in the process.

National prosperity is created, not inherited. Although national resources are a States' assets, the prosperity of a nation does not necessarily emerge solely from the natural endowments of the State concerned, nor from its labor resources, but rather from a certain localized process which engulfs economic structures, national values, culture and institutions. The essential catalyst to trade is national competitiveness.

National competitiveness is one of the most critical drivers of successful government and industry in every nation. Yet for all the discussion, debate, and writing on the topic, there is still no persuasive theory to explain national competitiveness. What is more, there is not even an accepted definition of the term "competitiveness" as applied to a nation. While the notion of a competitive company is clear, the notion of a competitive nation is not. The deliberations of the Conference clearly brought to bear this point and implicitly called upon States to revisit their own strategies with regard to the air transport industry on the basis that the most important feature of a competitive nation is its decisive characteristic that allows its companies to create and sustain competitive advantage in particular fields—the search is for the competitive advantage of nations. Of particular concern are the determinants of international success in technology and skill-intensive segments and industries, which underpin high and rising productivity.

Classical theory supports the principle that the success of nations in particular industries based on so-called factors of production such as land, labor, and natural resources is based on the fact that nations gain factor-based comparative advantage in industries that make intensive use of the factors they possess in abundance. Classical theory, however, has been overshadowed in advanced industries and economies by the globalization of competition and the power of technology.

Any new approach must recognize that in modern international competition, companies compete with global strategies involving not only trade but also foreign investment. What a new theory must explain is why a nation provides a favorable home base for companies that compete internationally. The home base is the nation in which the essential competitive advantages of the enterprise are created and sustained. It is where a company's strategy is set, where the core product and process technology is created and maintained, and where the most productive jobs and most advanced skills are located. The presence of the home base in a nation has the greatest positive influence on other linked domestic industries and leads to other benefits in the nation's economy. While the ownership of the company is often concentrated at the home base, the nationality of shareholders is secondary.

A new theory must move beyond comparative advantage to the competitive advantage of a nation. It must reflect a rich conception of competition that includes segmented markets, differentiated products, technology differences, and economies of scale. A new theory must go beyond cost and explain why companies from some nations are better than others at creating advantages based on quality, features, and new product innovation. A new theory must begin from the premise that competition is dynamic and evolving; it must answer the questions: Why do some companies based in some nations innovate more than others? Why do some nations provide an environment that enables companies to improve and innovate faster than foreign rivals?

In the continuing debate over the competitiveness of nations, no topic engenders more argument or creates less understanding than the role of the government. Is government an essential helper or supporter of industry, employing a host of policies to contribute directly to the competitive performance of strategic or target industries? Or is it the "free market" view that the operation of the economy should be left to the workings of the invisible hand.

Both views are seemingly incorrect. Either, followed to its logical outcome, would lead to the permanent erosion of a country's competitive capabilities. On one hand, advocates of government help for industry frequently propose policies that would actually hurt companies in the long run and only create the demand for more helping. On the other hand, advocates of a diminished government presence ignore the legitimate role that government plays in shaping the context and institutional structure surrounding companies and in creating an environment that stimulates companies to gain competitive advantage.

Government's proper role is as a catalyst and challenger; it is to encourage—or even push—companies to raise their aspirations and move to higher levels of competitive performance, even though this process may be inherently unpleasant and difficult.

In this context, the ICAO/IATA connection becomes even more important. IATA could provide ICAO with the necessary statistics and details that might justify such a request from a State, both before and during a possible dispute resolution situation before the Council. In this manner, ICAO can regain its lost involvement in developing air transport and justify its new Strategic Objective: *"Economic Development of Air Transport: Foster the development of a sound and economically-viable civil aviation system"*.

References

Banister D, Andersen B, Berechman J, Barrett S (1993) Access to facilities in a competitive transport market. Transport Plan Technol 17:341–348

Becker P (2002) Corporate foresight in Europe: a first overview. Institute of Science and Technology Studies, Bielefeld

Benjamin R (2012) Delivering our sustainable air transport future. ICAO J 67(6):5

Bisignani G (2006) Think again airlines, January 2006. At http://www.foreignpolicy.com/articles/2006/01/04/think_again_airlines?print=yes&hidecomments=yes&page=full

Bohmann K (2001) The ownership and control requirement in U.S. and European Union Law and U.S. Maritime Law – policy; consideration; comparison. J Air Law Commerce 66:690

Bouw P, Hall W (1999) Global competition in the airline industry. ICC World Bus Trade Rev 147–152

Mendelsohn AI (2012) International aviation today: what's wrong? Issues Aviat Law Pract 11 (3):333

Organization for Economic Cooperation and Development (1988) Deregulation and airline competition. OECD, Paris, p 1988

Oxford Economics (2009) Aviation, the real world wide web, June 2009. http://www.oxfordeconomics.com/free/pdfs/ox_econ_aviation_report/main.html

Porter ME (1980) Competitive strategy. The Free Press, New York

Chapter 3
Connectivity

It is disconcerting that neither the subject of connectivity nor the fundamentally important aspect of tourism was given even a passing mention at ATConf/5 in 2003.[1] The closest the Conference got to was in its Conclusions under the subject of "Consumer Protection" where it stated:

> As a premise in addressing consumer interests issues, States need to carefully examine what elements of consumer interests in service quality have adequately been dealt with by the current commercial practices of airlines (and service providers, if applicable) and what elements need to be handled by the regulatory and/or voluntary commitment approaches. States need to strike the right balance between voluntary commitments and regulatory measures, whenever the government intervention is considered necessary to improve service quality...[2]

3.1 The Declaration

The pivotal point and culmination of ATConf/5 lay in a *Declaration of Global Principles for the Liberalization of International Air Transport* adopted by the Conference by ovation, which, *inter alia*, emphasized the critical importance of safety and security in international air transport; noted the changes since the fourth Worldwide Air Transport Conference in 1994 in the regulatory and operating

---

[1] The 5th Worldwide Air Transport Conference of ICAO, under the theme *"Challenges and Opportunities of Liberalization"*, was held in Montreal from 24 to 28 March 2003 and attracted 794 participants from 145 Contracting States of ICAO and 27 observer Organizations. The Conference was held in the backdrop of an economic slowdown which had occurred in the two previous years, where the World's stock markets had fluctuated; growth slowed in Western Europe and other industrialized areas; the Japanese economy was bordering on recession and the United States and the rest of the world were still reeling from an unprecedented terrorist attack which had set the aviation world back in the amount of US$30 billion.

[2] Report of the Worldwide Air Transport Conference, Challenges and Opportunities of Liberalization, (Montreal, 24–28 March 2003) *ICAO Doc 9819; ATConf/5 2003* at 42.

R. Abeyratne, *Regulation of Air Transport*, DOI 10.1007/978-3-319-01041-0_3,
© Springer International Publishing Switzerland 2014

environment of international air transport brought about by economic development, globalization, liberalization and privatization; and reaffirmed the basic principles of sovereignty, fair and equal opportunity, non-discrimination, interdependence, harmonization and cooperation set out in the Chicago Convention which have served international air transport well and continue to provide the basis for future development of international civil aviation.

The *Declaration* called upon ICAO and its Contracting States, together with the air transport industry and other stakeholders in civil aviation, to work to ensure that international air transport continues to develop in a way that ensures high and improving levels of safety and security; promotes the effective and sustainable participation in and benefit from international air transport by all States, respecting national sovereignty and equality of opportunity; and takes into consideration the differing levels of economic development amongst States through maintenance of the principle of "community of interest" and the fostering of preferential measures for developing countries. The Declaration also calls for the providing of adequate supporting infrastructure at reasonable cost, facilitation of the provision of resources—particularly for developing countries, and allowing for growth on a basis that is economically sustainable, supported by adaptation of the regulatory and operating environment, in order to strive to limit its environmental impact. Also considered important is the meeting of reasonable expectations of customers and public service needs, particularly for low traffic or otherwise uneconomical routes, promoting efficiency and minimizes market distortions. Another important dimension is the call for a system which safeguards fair competition adequately and effectively; promotes cooperation and harmonization at the sub-regional, regional and global levels; and has due regard for the interests of all stakeholders, including air carriers and other operators, users, airports, communities, labour, and tourism and travel services providers; with the ultimate purpose of giving international air transport as much economic freedom as possible while respecting its specific characteristics and in particular the need to ensure high standards of safety, security and environmental protection.

One of the critical and thought provoking provisions in the Declaration is found in clause 4.4, which provides that each State will determine its own path and own pace of change in international air transport regulation, in a flexible way and using bilateral, sub-regional, regional, plurilateral or global avenues according to circumstances. Given that the overall approach of the Fifth Worldwide Air Transport Conference was "how to liberalize" (as against "whether to liberalize" which was the preoccupation of the earlier Air Transport Conference), this provision seems to say that the issue of how to liberalize is very much left to the States themselves, to be done at their own pace. If this be the purport and intent of this provision, it does not say anything much except to endorse what States had been doing prior to the Conference. One wonders in this context what the intent of the Contracting States of ICAO was, in adopting this platitude.

A repetition of the intent of this clause is found in the provision immediately proceeding clause 4.4, where States are requested that they should, to the extent feasible, liberalize international air transport market access, ensure air carrier access to international capital and air carrier freedom to conduct commercial activities—again a truism and fact of economic reality that had been happening in the aviation community before the Conference. Regarding cargo services liberalization, the same type of "endorsement" is seen in the statement that States should give consideration to liberalizing the regulatory treatment of international air cargo services on an accelerated basis, provided that clear responsibility and control of regulatory safety and security oversight is maintained.

Arguably, the most significant and thought provoking provision of the Declaration lies in clause 6.1, where it is acknowledged that ICAO should continue to exert the global leadership role in facilitating and coordinating the process of economic liberalization and ensuring the safety, security and environmental protection of international air transport. The importance of this statement lies in the fact that it has seemingly relegated ICAO's global leadership role in the process of liberalization to a position of mere facilitator and coordinator. Detractors of this provision could argue that it effectively precludes ICAO from attaining its objective, explicitly recognized in the Chicago Convention of "insuring" the safe and orderly growth of air transport and developing international air transport so as to meet the needs of the peoples of the world for safe, regular, efficient and economical air transport including preventing economic waste caused by unreasonable competition and insuring that the rights of Contracting States are fully respected and that every Contracting State has a fair and equal opportunity to operate international airlines. The argument could be further made that the profile of an Organization that has, by international treaty been given a leadership role of ensuring orderly growth of international air transport, along with insuring, *inter alia*, equality of opportunity for the operation of commercial air services has been undermined by assigning to it the role of facilitator and coordinator.

Such an argument, however, would be destitute of legal validity on the ground that fundamental principles of treaty law dictate that an unsigned *Declaration* would be subservient to an international treaty that is multilaterally signed and in force. In this sense, the Declaration would merely give a direction toward the goals at hand and leave the validity and authority of the provisions of the Chicago Convention untouched and unaffected.

3.2 The Tourism Connection

There are two main types of air travel: business travel and tourist or leisure travel (including what was called visiting friends and relatives travel). Connectivity[3] applies to both these categories but more so to tourism. At ATConf/6 the United

---

[3] While there is no universally agreed definition of "air connectivity", UNWTO follows a general understanding that it is an overall measure of the level of service—the range and economic importance of origins/destinations, the reliability and frequency of flights and connections—available through a country's aviation system linkage to the global air transport network. The higher the level of connectivity, the greater the level of access to the global economy. Facilitation

Nations World Tourism Organization (UNWTO)[4] argued that access by air is key
for both developed and developing countries and as such is air transport market
liberalization and that the on-going market access constraints are magnified when
they are translated to tourism (which includes both business and leisure visitors)
with substantial ramifications for major economies and small tourist destinations
alike. For many of the world's poorest countries, tourism is, or has the potential to
be, their major export; but without attractive air services the benefits of tourism for
these countries are limited.

The UNWTO argued that there should be a studied and positive approach to air
transport liberalization within the framework of the World Trade Organization and
its General Agreement on Trade in Services, which fully encompasses tourism but
not air transport. Advised the Conference that larger issues have effectively stymied
the development of revised trading agreements by the World Trade Organization
for a number of years and return to substantive address of the General Agreement
on Trade in Services (GATS) Annex on Air Transport Services was not envisioned
for at least some time. While this was disappointing, UNWTO was of the view that
it provided an opportunity for ICAO to develop its own construct on liberalization.
It suggested that a way forward might be for the ICAO to develop an air transport
regulatory framework along the lines of World Trade Organization provisions but
falling under the aegis of ICAO itself.

The suggestions of UNWTO at ATConf/6 were quite constructive. They
suggested that States base their assessment of international air service needs on
the basis of the broad economic and social benefits for their societies, including
tourism and trade as well as aviation factors and consider the application of the
Essential Service and Tourism Development Route concept developed jointly by
ICAO and UNWTO both as a safeguard and as an economic and social developmental tool. Another suggestion was that ICAO and UNWTO, both separately and
jointly, provide further guidance and actively promote the above approaches and
that ICAO actively promote liberalization of air carrier ownership and control, in
the first instance through endorsement and promulgation of the draft *Multilateral
Convention on Foreign Investment in Airlines*.[5] ICAO was requested in its consideration of modernizing the air transport regulatory framework, to take due account
of the provisions of the General Agreement on Trade in Services[6]; and that ICAO

---

measures in terms of passenger flow through airports are a condition of connectivity and, more
broadly, facilitation of visa processing (where required) is a necessary pre-requisite See A Tourism
Perspective on International Air Transport Association, ATConf/6-WP/63.

[4] UNWTO is the United Nations agency responsible for the promotion of responsible, sustainable
and universally accessible tourism. As the leading international organization in the field of
tourism, UNWTO promotes tourism as a driver of economic growth, inclusive development and
environmental sustainability and offers leadership and support to the sector in advancing knowledge and tourism policies worldwide.

[5] See Abeyratne (2012a) at 19–27.

[6] See Abeyratne (2012b), pp. 85–128.

and UNWTO position travel and tourism collectively as a strategic sector, with air transport as an interconnected core.

One of the shortcomings of ATConf/6 was that, although called the Sixth Air Transport Conference, which had within its remit all air transport issues, including those which fall within Annex 9 to the Chicago Convention on facilitation of air transport, this important subject was not addressed in the detail it should have. A bare passing mention was made under the subject of connectivity that Ministers of Tourism gathered in London from 4 to 8 November 2012 for the 6th UNWTO World Travel Market (WTM) Ministers' Summit and that the meeting concluded that complicated transit visa processes and policies that limit air connectivity continue to present major barriers to the growth of travel and tourism. Ministers and representatives from major tour operators and airlines further called for increased intra-governmental cooperation and support from the highest levels of government to break such barriers. Challenges related to transit visa processes in many countries remain a major obstacle to tourism development. Since more than 50 % of tourists are also air travellers, obstacles to tourism are obstacles to the development of air transport.

It is well accepted that transport is a vital and key driver of tourism as it forms a critical part of the tourism process. The current trend in tourism is the removal of entry barriers to many sectors of the tourism transport business through the challenge posed to large oligopolies by new entrants, essentially as a result of globalization. The low cost carrier[7] is a catalyst in this process. The secret of success of the low cost carrier is in the advantage it gains in being able to drastically lower its unit costs, unlike long haul carriers which are often burdened with a high cost base. However, in recent times, at least one long haul carrier—Condor—has introduced a low cost pricing structure on its intercontinental flights.[8] It is yet to be seen whether this experiment will be a success. Low cost carriers have, for several years, cut working capital and brought pressure on their asset bases, mostly by reducing their inventories and getting rid of uneven paying practices. This brings to bear the blatant difference between long haul legacy carriers, whose working capital is significant and low cost carriers whose asset base is comparatively low. For example, in the last decade (2003–2013) while the net operation of working capital of Alitalia was 8.1 % of its sales, and Air France retained a workable capital level of 2.2 % of sales, Ryanair in contrast showed a negative working capital of 2.8 % of its sales for the same year.[9]

Inherent to the metamorphosis was the radical change from a conventional commercial profile which existed from 1944 (when regulated international civil

---

[7] The low cost carrier model offers No frills (no meals offered on-board; one class of service; no lounges; no high-cost or complicated Frequent Flyer Programs; network structure, turnaround times (approx. 20–25 min.); lean distribution network (mainly call centers, Internet); E-ticketing; lower salary scales and high productivity (low employee-to-aircraft ratio); fleet commonality (one equipment type) and low training requirements; and operations to secondary airports.

[8] Condor Introduces Intercontinental LCC Pricing, *Aviation Daily*, Thursday, 27 May 2004 at p. 1.

[9] Korfman and Overeynder (2003–2004) at p. 28.

aviation was introduced) to the end of the twentieth century where a conventional purchasing structure existed permitting the purchase of air travel, anywhere in the world—seamlessly, to a system of air travel accessible through the relatively expeditious and simple means such as e-mail and the internet. Also, in the past, a complex set of relationships existed, called interlining, which offered the customer an infrastructure and procedures to be connected to a network of airlines. Aided by standards and recommended practices of ICAO and proactive regulations of IATA, customers who wished to travel internationally by air enjoyed low transaction costs and comfortable travel, ensured through a single call to a travel agent. These transactions ensured flexibility, refundability of fares in case of failure to travel, and often transferability to interlined carriers and others who were participating in the network of agreements that prevailed.

UNWTO further recorded that international tourist arrivals grew by 7 % in 2010 to a record 940 million, with positive growth reported in all world regions. Reflecting global economic trends, growth was driven largely by emerging economies, a development that looks set to continue over the coming years. The recovery of international tourism has confirmed the sector's extraordinary capacity to bounce back time and again from external shocks. Tourism is an extremely resilient sector and given its contribution to global economic growth, job creation and development, its faster-than-expected recovery in 2010 was welcome news. The global economic downturn in 2008–2009 has demonstrated more than ever the need for political recognition and support of the tourism sector. Throughout 2010, UNWTO worked to mainstream tourism in the global agenda by promoting its significant contribution to global prosperity, development and well-being. In 2010, UNWTO presented its Roadmap for Recovery—conveying the message that tourism means jobs, trade, economic growth and development—to leaders and decision.

UNWTO has reported that in 2011, international tourism receipts exceeded US $1 trillion for the first time, up from US$928 billion in 2010. In real terms, receipts grew by 3.8 %, following a 4.6 % increase in international tourist arrivals. An additional US$196 billion in receipts from international passenger transport brought total exports generated by international tourism in 2011 to US$1.2 trillion.[10]

---

[10] International tourist arrivals grew by close to 5 % during the first months of 2011, consolidating the 7 % rebound registered in 2010. According to the April Interim Update of the UNWTO World Tourism Barometer, growth was positive in all world (sub)regions during January and February 2011, with the exception of the Middle East and North Africa. South America and South Asia led growth (both at +15%), followed by Subsaharan Africa (+13 %) and Central and Eastern Europe (+12 %). Asia and the Pacific, the region with one of the fastest growth rates in 2010, saw its pace of growth slow down (+6 %), although from a very strong performance the previous year. Results were better than expected for Europe (+6 %), boosted by the recovery of Central and Eastern Europe, and the temporary redistribution of travel to destinations in Southern and Mediterranean Europe due to developments in North Africa −9 %) and the Middle East (−10 %). The Americas (+5 %) was in line with the world average, with strong results for South America and the Caribbean, but rather weaker growth in North and Central America.

According to the latest UNWTO World Tourism Barometer, international tourism receipts continued to recover from the losses of crisis year 2009 and hit new records in most destinations, reaching an estimated US$1,030 billion (740 billion euro) worldwide, up from US$928 billion (700 billion euro) in 2010. In real terms (adjusted for exchange rate fluctuations and inflation), international tourism receipts grew by 3.8 %, while international tourist arrivals increased by 4.6 % in 2011 to 982 million. This confirms the close correlation between both indicators, with growth of receipts tending to lag slightly behind growth of arrivals in times of economic constraints.

If tourism is to continue its trend of the last 2 years economic restraints, both on air transport and tourism, should have to be removed, or in the least minimized. The most compelling restraint pertains to the environment. Tourism involves travel for predominantly recreational or leisure involving the provision of services to support this leisure travel. Tourists are generally defined as people who "travel to and stay in places outside their usual environment for not more than one consecutive year for leisure, business and other purposes not related to the exercise of an activity remunerated from within the place visited". Tourism has become a popular global leisure activity.

UNWTO recorded in 2007 that the number of international tourist arrivals has risen from 25 million in 1950 to 840 million in 2006. The revenues generated by these arrivals—not including airline ticket sales and revenues from domestic tourism—have risen at an average rate of 11.2 % a year (adjusted for inflation) over the same span of time, which is nearly twice as fast as arrivals and a growth rate that far outstrips that of the world economy as a whole. International tourism receipts reached $735 billion in 2006, almost 900 billion including air tickets, making tourism not only a socio-economic driver but one of the largest categories of international trade. Tourism represents one quarter of all exports of services—40 % with air transport revenues included. It is also noteworthy that the growth trend will continue, as according to a news release of 18 September 2007 issued by the International Civil Aviation Organization (ICAO), total world airline scheduled passenger traffic in terms of passenger-kilometers is expected to grow at an average annual rate of 4.6 % up to the year 2025, half a percent point lower than the growth rate achieved over the period 1985–2005, according to forecasts prepared by ICAO. Total freight traffic growth over the same period is forecast to be stronger, at 6.6 % per annum in terms of freight tonne-kilometers.

There can be particular benefits from tourism for the poorest economies, where international arrivals are growing at twice the rate in the industrialized States. Although they are still at a low level, the tourism receipts of the Least Developed Countries increased fivefold between 1990 and 2005, from $1 billion to $5 billion. Tourism has become one of the largest sources of foreign exchange revenues for developing countries generally and for the 49 LDCs specifically, reducing their foreign debt and diversifying their economies. Tourism is often the principal service sector activity and it is a notably effective catalyst for gender equality, employment of young people, rural regeneration, cultural preservation and nature conservation.

Therefore UNWTO, which is the specialized agency of the United Nations with a central and decisive role in promoting the development of responsible, sustainable and universally accessible tourism, concludes that, for all these reasons, tourism can play a major role in improving the standard of living of disadvantaged populations and helping them lift themselves above the poverty threshold. Tourism can be a primary tool for achievement of the Millennium Development Goals of the United Nations, as long as a balance with climate change effects is maintained.

Tourism—business and leisure travel—is a significant global industry making positive contributions to growth, trade and development, with particular potential for poor countries. Tourism is both affected fundamentally by climate change and a contributor to the global emissions of greenhouse gases with air transport being a substantial contributor to total tourism emissions.

There is no doubt that tourism and air transport are symbiotic. Travel and tourism, the largest combination of industries and the largest creator of wealth, is estimated to generate $3.5 trillion a year in activity and potentially provides employment to 130 million people worldwide. This accounts for 10 % of the world's GDP, 10.3 % of the world's wages, 9.8 % of the profits and 11.7 % of indirect and direct taxes. WTO has recorded that about 40 % of the 840 million international tourist arrivals in 2006 were by air. In terms of long haul destinations, this figure could be even higher. Furthermore, the vast majority of the 931 million international passengers in 2006 were tourists.

Therefore, it is only logical that tourism and aviation together should actively involve the international community on issues of climate change mitigation. At a recent conference conducted by ICAO and McGill University in Montreal, UNWTO stated that Green House Gas emissions from tourism are estimated to contribute about 5 % in terms of global carbon dioxide emissions and 4.6 % in terms of radiative forcing. WTO also observed that air transport accounts for an estimated 40 % of the tourism contribution of carbon dioxide and 54–75 % of the radiative forcing. Furthermore air transport accounts for an estimated 60 % of the international tourism contribution of carbon dioxide and is overwhelmingly dominant at medium and long haul air transport operations.

UNWTO claims that climate change is not an abstract concept for tourism. It is a phenomenon which already affects the sector and certain destinations in particular: most products directly incorporate some conditions that are vulnerable to climate change—winter sports, sun-and-sea, eco-tourism, are all at risk. Many destinations are vulnerable—rising water levels will affect low lying coastal zones and small islands; snow line shifts will affect mountain resorts; desertification and deforestation will affect ecotourism sites; and temperature change will create serious impacts on traditional mass market venues.

Although the logical solution to this issue might lie in a multitude of options to reduce emissions in the tourism industry, by far the greatest potential offered lies in the air transport sector. The sustainability of tourism could be profoundly affected by the reduction of the number of air operations and flight distances. Other areas that might facilitate sustainable development of tourism through the aviation sector would be in the advancement of airframe and engine technology and efficiency gains through effective air traffic management and operational practices of airliners. The use of alternative modes of transport and communications wherever possible would also be a plus factor. Economic instruments such as emissions trading and carbon offsetting are also positive measures that would curb aircraft engine emissions. There must be global enforcement of these measures. Also, alternative fuels should be considered a necessary alternative to fossil fuels.

At the present time, however, market conditions have changed drastically and air transportation is growing twice as much as the general economy.[11] The result of this trend is a dramatic increase in the size of aggregate and individual aviation markets, resulting in turn in the emergence of the new breed of air carrier—the low cost carrier—which offers a simple and consequently low-cost service aimed at attracting customers with simple itineraries. The low cost carrier has grown to such significant lengths so as to compete with the largest established carriers in the world.

3.3 Airport Slots

Connectivity, especially through hub operations, would depend largely on the allocation of airport slots[12] for connecting flights. At ATConf/6. IATA stated in its report that the total number of capacity constrained airports that have been labelled as a fully coordinated or Level 3 Airport[13] subject to slot allocation under the IATA Schedule Coordination System continues to increase. There were 136 in 2000, 155 in 2010, and by 2012, the number is expected to reach 159 (104 in Europe, of which 92 are in the 27 EU Member States, 43 in Asia Pacific, and the remaining 11 scattered in the Middle East, North America, and South Africa).

---

[11] Tretheway (2004), pp. 3–14 at p. 4.

[12] Airport slots are specific time periods allotted for an aircraft to land or take off at an airport. Where the demand for slots at a particular airport exceeds the available supply, the airport can be considered "capacity-constrained", at which time a "slot allocation" process is implemented. Capacity constraint may occur only at certain periods of the day or on certain days of the week, or even during specific seasons. An airport slot is a designated day and time usually within a 15 or 30 min period for an aircraft to arrive at or depart from an airport. See *Regulatory Implications of the Allocation of Flight Departure and Arrival Slots at International Airports*, ICAO Doc 283-AT/ 119, 2001 at Chapter 1, p. 1. European Council Regulation (EEC) No 95/93 of 18 January 1993 on common rules for the allocation of slots at Community airports defines a slot as "the scheduled time of arrival or departure available or allocated to an aircraft movement on a specific date at an airport co-ordinated under the terms of the Regulation".

[13] Level-3 airports are airports where the demand for runway and gate access exceeds the capacity of the airport, resulting in the need for slots to be allocated to airlines through the IATA Schedule Coordination System. They are airports where capacity providers have not developed sufficient infrastructure, or where governments have imposed conditions that make it impossible to meet demand. A coordinator is appointed to allocate slots to airlines and other aircraft operators using or planning to use the airport as a means of managing available capacity. See IATA Worldwide Slot Guidelines Effective January 2013, Fourth edition, 10.

In addition, 121 airports across the world are experiencing some level of congestion. If traffic volumes continue to increase at a pace faster than investment in capacity expansion, it is expected that many of these 121 airports experiencing congestion will become fully-coordinated or Level 3 airports.

There are several criteria for the allocation of slots: slots are only allocated for planning purposes by a duly appointed coordinator at a Level 3 airport. Slots are only allocated to airlines and other aircraft operators. An airline or other aircraft operator must have a slot allocated to it before operating at a Level 3 airport. Certain types of flights (for example, humanitarian or state flights) may be exempt or subject to special local procedures. Airlines and other aircraft operators must not intentionally operate services at a significantly different time or use slots in a significantly different way than allocated by the coordinator A series of slots is at least five slots requested for the same time on the same day-of-the-week, distributed regularly in the same season, and allocated in that way or, if that is not possible, allocated at approximately the same time. An airline is entitled to retain a series of slots on the basis of historic precedence. Historic precedence applies to a series of slots that was operated at least 80 % of the time during the period allocated in the previous equivalent season. Historic slots may not be withdrawn from an airline to accommodate new entrants or any other category of aircraft operator. Confiscation of slots for any reason other than proven, intentional slot misuse is not permitted. Slots may be freely transferred or exchanged between airlines, or used as part of a shared operation, subject to the provisions of these guidelines and applicable regulations. Coordinators must be functionally and financially independent of any single interested party and act in a neutral, transparent and non-discriminatory way. The allocation of slots is independent from the assignment of traffic rights under bilateral air service agreements.

Since ATConf/5, the issue of airport capacity constraints and the search for appropriate solutions to the slot allocation problem continue to be a challenge for governments, airport operators and the airline industry. States' policies and practices in dealing with the slot allocation issue also vary. Most countries have been applying the IATA system which is presented in the IATA "Worldwide Slot Guidelines (WSG)"; these guidelines contain a set of rules agreed upon by airlines and slot coordinators. Some States have adopted their own regulations, often drawing on the principles in the IATA WSG. For example, the European Union (EU) and the United States have enacted their own regulations on slot allocation, which have undergone some revisions or adjustments during the past few years. In some cases, States have addressed the issue within the context of their bilateral ASAs. There are also instances where "slot trading" has been allowed at highly congested airports. Such fragmentation of regulatory approaches has drawn some criticism in the recent years.

In addition, it has been pointed out that the existing slot allocation procedures lack the sanction mechanisms which could disincentivize some air operators to adopt practices that decrease the efficiency of the slot allocation process, such as overbidding of slots, late return of slots, and under-utilization. These deficiencies need to be addressed.

It is clear that insufficient airport capacity and slots have a negative impact on the ability of air carriers to exercise market access rights, but it is a physical impediment that cannot be resolved with short-term solutions. Due to the nature of the problem, it is difficult to prescribe global solution, as capacity shortages vary from airport to airport. However, there is a need for States, air carriers and airport operators to address the issue with a broader and long-term perspective, giving due regard to the capacity requirements and constraints in the planning of infrastructure development. In this regard, at the recommendation of ATRP, the Secretariat conducted a second survey on present and future airport capacity constraints and provides information on the survey results at Appendix B. It should be noted that the survey results are not sufficiently conclusive to provide any relevant information to the Conference. With respect to difficulties or disputes over slot allocation, the most practical approach for States to address the disputes as recommended by the ICAO guidance, i.e., through the consultation and dispute resolution mechanisms under ASAs or through other available fora. In the meantime, there is a need for States, in coordination with the aviation industry and ICAO, to continue to review the implementation of existing slot rules and procedures such as "use-it or lose-it," slot trading, slot buying/selling, and local rules at variance with the IATA WSG, and consider whether changes are appropriate, considering however the need for global compatibility of rules. In this regard, ICAO can continue to play an important role in developing guidance and/or exploring appropriate ways by which to address the issues, taking into account the interests of States, the industry, and other stakeholders.

At ATConf/6 ICAO advised the Conference that the issue of slot allocation is linked to specific local situations, but affects market access and operation of international air services. As air traffic continues to grow and hub-and-spoke operations continue to increase, the issues will remain and, potentially, grow. As the situation varies from State to State and airport to airport, it is difficult to prescribe a global solution. However, there is a need for the international aviation community to align existing slot allocation rules and procedures, and to explore how States and aviation stakeholders might work together towards improved policy approaches:

(a) The most practical method to resolve specific difficulties is through consultation at the local level between the parties concerned, taking into account obligations under relevant international agreements, applicable national and regional rules, and the interests of all stakeholders;

(b) ICAO has addressed the issue of slot allocation extensively and has developed related guidance for use by States, which remains relevant; and

(c) Based on the results of the surveys, and it spite of the close cooperation undertaken with and by the industry, it appears that providing meaningful data and forecasts on the present and future capacity constraints of airports is extremely difficult, due to the lack of available information.

In view of the above, ICAO recommended that States should give due consideration to the concerns of other States over the issues related to slot allocation and the negative impact on international air services and make every effort to resolve the problems. States should also give due consideration to capacity demands in the planning and development of aviation infrastructure. In this context, ICAO should continue to monitor the situation and States' practices in handling the issues of slot allocation, raise State awareness of ICAO policy guidance, and encourage its use by States. It was also recommended that ICAO should continue to play a leadership role in developing policy guidance and work in close cooperation with both States and aviation stakeholders in exploring appropriate ways by which to address market access and slot allocation issues, taking into account the interests of States, the industry, and other aviation stakeholders and also keep States informed of any significant developments, and ICAO future work in this regard, including information on airport and ATM capacity demands and/or constraints. The Arab Civil Aviation commission (ACAC) requested that ICAO adopt policy recommendations that ensure that carriers from developing countries are granted preferential treatment in the allocation of slots.[14]

Slot allocation and connectivity are interlinked and have a definite profitability dimension for the airlines and a distinct connotation of flexibility of choice for the consumer. In addition to the restrictions placed on the free movement of aircraft between States on a commercial basis, airlines face a further hurdle when it comes to obtaining airport slots once they obtain the right to carry traffic in and out of a State. Slots are allocated to aircraft operations in an order of priority as follows: regular scheduled services; ad-hoc services; and other operations. Increasing congestion at major airports brings to bear the increasing difficulties faced by air carriers to obtain slots to land at and depart from airports. Airlines are finding new and innovative ways to bypass the restrictions imposed by Article 6 of the Chicago Convention through various commercial tools. However, there are still the remnants of an unhealthy competition between dominant carriers who retain their larger market shares against carriers who do not have as large a competitive profile as their opponents.

The issue of lack of slots at capacity-constrained airports is not new. Over the past decade, growth in air traffic has continued to outstrip available capacity at many airports. According to IATA, the total number of fully-coordinated airports subject to slot allocation under the IATA Schedule Coordination System had increased from 136 in 2000 to 142 in 2008 (Europe has an increase of 16 %). Difficulty in obtaining slots can affect the ability of an air carrier to exercise its market access rights granted under relevant bilateral air services agreements.

All formal mechanisms for dealing with the lack of airport capacity are based on the concept of an airport slot, which is the time that an aircraft is expected to arrive at or depart from a capacity-constrained airport. For commercial operations which use airport gates, this time is calculated based on when the aircraft arrives at or leaves the gate. To take into account variations in flight times, unavoidable delays, etc., airport slots may actually be allotted in terms of a time period, such as 16:45 to

---

[14] Allocation of Slots and Their Impact on Air Transport, *ATConf/6-WP/25*, at 2.

17:00. Airline schedules, of course, are stated in more precise terms and, for example, five flight arrivals in that time period could each appear in the respective airlines' schedules at a specific time, for example 16:45, 16:48, 16:52, 16:53 and 16:58.

However, different airlines may schedule their flight departures at the same time (for commercial or operational reasons), for example on the hour, which at busy airports can exacerbate peaking and often result in aircraft having to wait in line for a take-off clearance. An airport slot should not be confused with an air traffic control (ATC) slot, the take-off or landing time of an aircraft which is assigned by the relevant ATC authority to make optimum use of available capacity at points en route or at the destination airport by sequencing the air traffic to regulate its flow efficiently. Thus, commercial operations may not land or take off in the same order as reflected in their respective schedules, but at times which would enable air traffic control to regulate efficiently the flow of aircraft into or out of the airport and the en-route system. This may involve, for example, interspersing commercial flights with general aviation flights and varying the order of take-off or landing to take account of greater separation requirements for larger aircraft, late arriving aircraft, etc. With the assignment of an airport slot, airlines can build their schedules, taking into account time to taxi to and from gates and customary en-route time, on the assumption that an ATC slot will be made available as close as possible to the time necessary for the flight to operate on schedule. This underlies the importance of close coordination between the coordinator assigning the airport slots and the air traffic control authorities responsible for flowing aircraft into take-off, landing and the en-route system.

An airport slot is essential in order to mount commercial services at an airport which has a slot allocation regime, but it is also part of a multifaceted package of services and facilities provided by different entities, such as gates, air traffic control, ground handling, passenger and cargo processing—all of which require close coordination and cooperation between and among national authorities, airports and airlines In a few slot allocation regimes, procedures differ depending on the type of entity using them. Thus, there can be commuter slots, air carrier slots, new entrant (to a city-pair market) slots, and slots for general aviation, military, domestic or international flights.

International access to airports is governed, inter alia, by Article 15 of the Chicago Convention, the first sentence of which provides that every airport in a contracting State which is open to public use by its national aircraft shall likewise, subject to the provisions of Article 68, be open under uniform conditions to the aircraft of all the other contracting States.

Article 15 establishes a national treatment standard for all contracting States in the context of the use of airports and other air navigation facilities for international air transport. Insofar as the operation of scheduled services is concerned, Article 15 subjects the use of airports to Article 68, which permits each contracting State, subject to the provisions of the Convention, to "designate the route to be followed within its territory by any international air service and the airports which any such service may use". However, Article 68 should not be read individually so as to derogate the national treatment standard established in Article 15.

Article 15 does not itself accord a right to operate international scheduled or non-scheduled air services. The operation of international scheduled services is subject to Article 6 of the Convention, which provides that "no scheduled international air service may be operated over or into the territory of a contracting State, except with the special permission or other authorization of that State, and in accordance with the terms of such permission or authorization." With respect to international non-scheduled operations, although Article 5 of the Convention allows aircraft engaged in non-scheduled commercial flights to overfly or make non-traffic stops in the territory of a Member State without the necessity of obtaining prior permission, authorization is generally required for such commercial flights from any State in which passengers and/or cargo are loaded or unloaded.

The authorization required for international scheduled air services in Article 6 is customarily accorded on the basis of traffic rights exchanged bilaterally or regionally, which either name a specific city at which such rights may be exercised or make a broader, non-specific grant in terms of any city in a State's territory. Although some bilateral and regional agreements include provisions for non-scheduled flights, the general practice has been for the States concerned to approve such flights on the basis of national regulations and policies. Regardless of the underlying source of the authorization to operate international commercial air services, once granted, national treatment and the uniform condition criteria of Article 15 apply.

A combination of this clear standard of treatment and the practice of having common traffic points for national and foreign airlines may explain why bilateral agreements, except in rare instances, do not deal with slot allocation or access to specific airports. However, air service negotiators have to be mindful of the difficulties of obtaining access to capacity-constrained airports and take into account that the rights they are seeking for their airlines at those airports may not be able to be exercised for some time.

The uniform treatment principle is important also with respect to conditions on the use of airports, particularly for environmental purposes. Thus curfews, or aircraft noise criteria, as well as any exceptions thereto, must be applied uniformly to both national and non-national aircraft engaged in similar international services. Inter-governmental disputes involving airport access under the uniform treatment rule have been rare; it has been more common for airport access disputes to focus on specific cases where airlines which have the underlying route rights to serve a city have not been able to secure access or increase service to that city's airport because of a lack of available slots. In such cases, States have usually relied on the bilateral provision which requires that designated air carriers have a fair and equal opportunity to operate or compete with respect to the services covered by the agreement.

There has been a continuing debate as to the "ownership" of airport slots, primarily in terms of claims by airlines which have historically used them for long periods of time. However, some formal regulatory regimes either explicitly or implicitly exclude this concept, for example, stating that airlines do not acquire property rights to the slots assigned to them and that the slots must be returned to the aeronautical authority under certain circumstances. The implicit approach ties the continued use of the slot to its use at a specified level (e.g. 80 %) and allows the exchange of slots on a one-for-one basis. In one instance in the United Kingdom, a court ruled in March 1999 that financial considerations in connection with an exchange of slots under the European Union (EU) common slot rules did not invalidate the exchange. However, the court did not rule on whether the exchange as such involved real property.[15]

Nevertheless, the obvious value in terms of market access of slots at airports with severe constraints on capacity has led to the treatment of these slots as a de facto financial asset of the airlines holding them. Thus, the purchase by one airline of another will take into account an estimated value of the airport slots involved. However, regulatory authorities have retained the right to approve or disapprove the transfer of airport slots in this manner, primarily through the approval or disapproval of the purchase or merger involved. The only formal pricing of airport slots has occurred in the United States where the purchase, sale and lease of certain domestic slots at the four airports currently subject to the Federal Aviation Administration's High Density Rule has been permitted since 1986. This has led some airlines serving these airports which purchase such slots to reflect their value as assets in their financial accounts.

The slot allocation issue has been extensively addressed by ICAO. In 1999, a detailed study on it was conducted by the Secretariat and submitted for review by the Conference on the Economics of Airports and Air Navigation Services (ANSConf2000) held in 2000. The study, which was published in 2001,[16] analyzed the trends for airports where the demand exceeds capacity supply; the regulatory framework involved; and the means by which governments, airports and airlines have sought to alleviate this problem. It also assessed existing and potential mechanisms for dealing with a chronic shortage of airport capacity and suggested possible improvements of, and alternatives to, the existing systems.

ATConf/5 addressed the issue of airport constraints in the context of market access. Recognizing that the ability of air carriers to exercise market access rights granted under relevant air services agreements is closely linked to the availability of slots at the airports concerned, the Conference concluded that "in liberalizing market access, due consideration should be given to airport capacity constraints and long-term infrastructure needs. Problems involving air carriers which are unable to exercise their entitled traffic rights at a capacity-constrained airport

---

[15] In this case, it was reported that Britain's number two long-haul airline, Virgin Atlantic Airways, and Australia's flag carrier Qantas each paid an alleged 20 million GBP to the small British regional airline Flybe for six slots at Europe's biggest airport, London Heathrow. Heathrow is known as one of most—if not the most—slot constrained airports in the world, with airlines queuing for years to gain access to the airport. See The Sunday Times, Heathrow Slots Take Off, January 4, 2004; *O'Connell*, The Sunday Times, Soaring Costs Of Touching Down, 22 February 2004.

[16] Circular, 283, *infra*, note 22.

may, if necessary, be addressed in the context of discussions on the relevant air services agreements. In this regard, sympathetic consideration should be given to the request for preferential treatment from those States whose airports are not slot-constrained but whose air carriers are unsuccessful in obtaining slots at slot-constrained airports, consistent with relevant national legislation and international obligations." The Conference further concluded that "any slot allocation system should be fair, non-discriminatory and transparent, and should take into account the interests of all stakeholders. It should also be globally compatible, aimed at maximizing effective use of airport capacity, simple, practicable and economically sustainable."

As recommended by the ICAO guidance, the most practicable measure States can take in addressing a slot problem is through relevant consultation or dispute settlement mechanism under the bilateral air services agreements, in accordance with the principles of fair and equal opportunity and reciprocity and in a spirit of cooperation and mutual understanding. However, the reality is that some States do not follow or implement ICAO's guidance. This may be largely due to the fact that ICAO's policies and guidance in the economic field, unlike the Standards and Recommended Practices (SARPs) in the technical field which are binding for States, are of a recommendatory nature, for optional use by States. The lack of application of ICAO's policies and guidance is not conducive to the general interest of the Organization and its member States. The question as to how to improve this situation involves a broader policy issue, which will need to be considered by the Council and States at an appropriate time and forum.

Problems of slot availability are mostly caused by congestion at airports and they should be addressed in the most efficient manner. Eliminating exemptions on charges for smaller airlines or airlines that operate fewer frequencies to an airport, and curbing airport monopolies are effective measures to address the congestion issue.[17] One of the ways in which airlines obviate the problem of unavailability of slots is through alliances with each other. Airline alliances are usually created with the intention of putting into place an integrated network of products, services and standards between two or more carriers who have the objective of operating air services more efficiently by eliminating burdensome duplication of costs and achieving better services for the travelling public.[18] These alliances have also been identified as "a distinct form of entry mode that has been used as a low-cost means of gaining access to new markets and local infrastructure".[19] The underlying philosophy of the airline alliances, is not so much an emphasis on the more effective use of resources such as labour, capital and national resources (which are inevitably important factors) but rather an overall reliance on the strategy of location, where the sharing of locations represented by the various airlines have enabled them to

---

[17] Levine (2009), p. 37 at 76–87.

[18] Kimpel (1997), pp. 475–513 at 476.

[19] Doz et al. (1990), pp. 17–143 at 33. For a detailed categorization of alliances see Rhoades and Lush (1997), pp. 109–114.

produce their goods and services in a consistent manner, thus achieving the status equivalent to a cartel, while still retaining their individual identities. One of the advantages of forming an alliance is that airlines could share their slots within such an alliance to maximize operations and revenue.

3.3.1 Night Curfews

Slot allocation is severely affected by nigh curfews imposed by some States to prevent aircraft landing and taking off during night time. Airport night flight restrictions or curfews are rules imposed on aircraft operators that prohibit aircraft take-offs and/or landings during a specified period of time. Such night flight restrictions may apply to all aircraft or only to certain aircraft, according to their noise performance. Most restrictions are adopted as a measure by which to address the adverse effects of aircraft noise on the affected airport and nearby communities. While the measure of night flight restrictions can help reduce the aircraft noise problem at the airport, it can also have an impact on the operation of air services, particularly international services to/from the airport, and the economic well-being of the local community and the country at large.

The issues arising from such measures have existed for many years and continue to remain despite the fact that advancement in aircraft engine technology has contributed considerably to noise abatement in the past two decades. Due to continuous traffic growth and due to increased pressure from airport neighbours, there is growing pressure to impose operating restrictions on night flights in some regions of the world. The pressure is quite intense for some major hubs and, in some cases, at secondary airports which are located in very densely populated areas. In many instances, inadequate land-use management policies have allowed urban encroachment around airports, resulting in an increase of the number of people significantly exposed to aircraft noise in spite of an actual reduction in noise emissions. Furthermore, conditions for allowing airport expansion sometimes requires a strong commitment from both airports and aircraft operators to limit or reduce the general noise level. In addition, as the number of flights increase, the population in the vicinity of the airport often becomes more concerned with health problems, including those caused by aircraft noise. For instance, it has been reported that noise can be one of many environmental stressors, and there is evidence that acute noise exposure can cause temporary elevations in heart rate. Of note is the fact that the reported level of annoyance from aircraft noise is often increased by factors that are not related to noise, such as congestion due to road traffic in the vicinity of airports, the fear of air accidents, or financial concerns about the value of property around the airport.

ICAO has made regular attempts at addressing the night curfew issue. In response to concerns voiced by Member States, ICAO has addressed the issue of night curfews. As early as in 1989, ICAO, in its Assembly Resolution A27-11 (still in force), invited States "to consider the possible relaxation of operating restrictions for aircraft meeting the requirements of Chapter 3 of Annex 16 — *Environmental Protection*, including the easing of night curfews and/or quotas for off-schedule arrivals by such aircraft". ATConf/5 also addressed the issue in the context of market access. It was noted that the abolishment of night curfews would increase airport capacity and assist in resolving the problem of airlines unable to exercise traffic rights at certain airports. However, this would create difficult environmental and political problems at the airports concerned.

The Conference was requested to consider the possibility of States giving due consideration to the concerns of other States and the negative impact on international air services when dealing with issues of night flight restrictions and make every effort to resolve the problems with States concerned through consultation and available dispute settlement mechanisms. States were also asked to respect and follow the ICAO Balanced Approach[20] principle in their regulatory action on aircraft noise management at airports, giving due regard to the views of all stakeholders, examining alternative means of addressing the problems, evaluating the likely costs and benefits of various measures and striving for the most cost-effective solutions.

ICAO was also requested to continue to play a leadership role in developing policy guidance and should work in close cooperation with States and the industry to explore appropriate ways to address slot allocation issues, including possible new avenues, taking into account the interests of States, the industry, and other aviation stakeholders, and continue to monitor States' practices in handling this issue and keep States informed of any significant developments. ICAO should also raise State awareness of ICAO policy guidance and encourage its use by States.[21]

Although the situation in each capacity-constrained airport varies widely, with the steady increase in air traffic it is clear that more and more States will be confronted with slot allocation decisions. These decisions will be particularly difficult in the international arena because, as a market access issue, slot allocation involves which airlines will operate to and from a capacity-constrained airport and which will not, often when airlines from both groups have the underlying traffic rights and authorization to operate international air services to and from the city in which the airport concerned is located.

Increasing airport capacity through new or enlarged airports, runways and terminals is clearly the best solution for a capacity-constrained airport. However, it is equally clear that this solution is not feasible at a number of airports with environmental, physical and other constraints which prevent their replacement or

---

[20] To address the aircraft noise problem, the ICAO Assembly has endorsed a "Balanced Approach" to noise management (Assembly Resolution A37-18, Appendix D; Doc 9829, *Guidance on the Balanced Approach to Aircraft Noise Management*). The approach "consists of identifying the noise problem at an airport and then analysing the various measures available to reduce noise through the exploration of four principal elements, namely, reduction at source, land-use planning and management, noise abatement operational procedures, and operating restrictions, with the goal of addressing the noise problem in the most cost-effective manner".
[21] Night Flight Restrictions, *ATConf/6-WP 8*.

expansion. For these cases, States must find some means to deal effectively and fairly with situations where the demand to operate commercial air services exceeds the capacity of the airport.

Measures to manage a lack of airport capacity can improve the situation in the short term and help to avoid bilateral disputes related to the allocation of airport slots for international services. Improvements in air traffic control and groundside facilitation as well as to existing mechanisms for slot allocation can increase the use of existing capacity and thereby provide some relief from a shortage of airport capacity.

Some issues related to capacity-constrained airports will involve broader regulatory policy questions, such as the enhancement of competition, the avoidance of excessive concentration and abuses of dominant positions, as well as the compatibility of broad market access with capacity-constrained airports. Although the broad granting of traffic rights bilaterally and regionally with multiple airline designation creates additional potential demand for airport slots, it also provides some relief in the form of flexibility to use alternate airports and cities which can accommodate new and increased air services.

A number of States will nevertheless have the task, in the long term as well as the short term, of balancing conflicting objectives in terms of which international air services will be able to use their capacity constrained airports. In fashioning responses to this problem, States will have to take into account the legal framework provided by the Chicago Convention, air services agreements, regional and national slot allocation rules and existing voluntary mechanisms for managing insufficient airport capacity. However, the response will have to fit the situation of the individual airport(s) concerned and will therefore vary depending on the nature of the constraint and the means taken to overcome it.

There is ICAO policy on slot allocation.[22] A more functional set of policy guidelines is contained in the IATA Worldwide Slot Guidelines (WSG) which is a set of standards and best practices developed by IATA Member airlines and the airport coordinator community.[23] These guidelines are a comprehensive set of procedures for the allocation and management of airport capacity. The principal users of these guidelines are airlines and airport coordinators. IATA also conducts the IATA Slot Conference (SC) which is the forum for the coordination of planned operations at Level 2 and Level 3 airports,[24] held twice each year for the summer

---

[22] Regulatory Implications of the Allocation of Flight Departure and Arrival Slots at International Airports, *Circular 283-AT/119.*

[23] The Worldwide Slot Guidelines (WSG) is a set of standards and best practices developed by IATA Member airlines and the airport coordinator community. These guidelines are a comprehensive set of procedures for the allocation and management of airport capacity. The principal users of these guidelines are airlines and airport coordinators.

[24] Level 1 airports are airports where the capacity of the airport infrastructure is generally adequate to meet the demands of airport users at all times. Level 2 airports are airports where there is potential for congestion during some periods of the day, week, or season which can be resolved by voluntary cooperation between airlines. A facilitator is appointed to facilitate the planned operations of airlines using or planning to use the airport.

and winter seasons. The June SC addresses the following winter season and the November SC addresses the following summer season. WSG requires that the airport managing body or other competent authority should regularly conduct a thorough demand and capacity analysis, using commonly recognized methods. In particular, demand and capacity should be assessed whenever there are significant changes in airport infrastructure, operational practices, or patterns of demand.

The analysis should objectively consider the ability of the airport infrastructure to accommodate demand at desired levels of service, such as queue times, levels of congestion or delay. The analysis should assume that the airport facilities are being managed efficiently and are fully staffed. The analysis should also determine any infrastructure, operational, or environmental constraints that prevent demand being satisfied and identify options for overcoming such shortages through infrastructure, operational or policy changes and improvements. The results of the demand and capacity analysis should be made available to interested parties in order to encourage cooperation and to alleviate and resolve any constraints on demand.

References

Abeyratne R (2012a) Aeronomics and law: fixing anomalies. Springer, Heidelberg

Abeyratne R (2012b) Administering the skies – facing the challenges of market economics. Aracne Editrice, Rome, pp 85–128

Doz Y, Prahalad CK, Hamel G (1990) Control, change and flexibility: the dilemma of transnational collaboration. In: Bartlett C, Doz Y, Hedlund G (eds) Managing the global firm. Routledge, London, pp 17–143

Kimpel S (1997) Antitrust considerations in international airline alliances. J Air Law Commerce 63:475–513

Korfman R, Overeynder W (2003–2004) Liberation of capital. Airlines international, IATA 12/1: 28

Levine ME (2009) Airport congestion: when theory meets reality, law and economics research paper series, working paper no. 08-55. Yale J Regul 26(1):37

Rhoades DL, Lush H (1997) A typology of strategic alliances in the airline industry: propositions for stability and duration. J Air Transport Manag 3(3):109–114

Tretheway MW (2004) Distortions of airline revenues: why the network airline business model is broken. J Air Transport Manag 10(1):3–14

Chapter 4
The Facilitation Connection

It is a curious fact that no one in particular at ATConf/6 pushed facilitation as a catalyst that could effectively promote tourism and develop air transport. The ICAO Secretariat made a feeble attempt by entitling a working paper with the title "facilitation"[1] where there was no mention of the value of an e-Passport as a tool that facilitates and enhances air travel, no mention of the recommendation to do away with visas as much as possible and certainly no mention of Annex 9[2] to the Chicago Convention on facilitation which has several compelling provisions in this regard. The paper, in passing, and as a half-hearted attempt, recommended that States should cooperate in the identification of main impediments to air transport connectivity; and that ICAO should undertake a cost–benefit analysis of measures aimed at enhancing connectivity, including those of facilitation.

Annex 9 has, as its theme, the elimination of delays in air transport and the mitigation of the adverse effects of bureaucracy, both of which would assist in optimizing economic development, connectivity and cost reduction. The Annex accomplishes these goals through Standards[3] and Recommended Practices.[4] The Standards and Recommended Practices on Facilitation are the outcome of Article 37 of the Convention, which provides, *inter alia*, that the "International Civil Aviation Organization shall adopt and amend from time to time, as may be

---

[1] ATConf/6-WP/20.

[2] Annex 9 to the Convention on International Civil Aviation, 12th Edition, 2006.

[3] Annex 9 defines a Standard as any specification, the uniform observance of which has been recognized as practicable and as necessary to facilitate and improve some aspect of international air navigation, which has been adopted by the Council pursuant to Article 54 (*l*) of the Convention, and in respect of which non-compliance must be notified by Contracting States to the Council in accordance with Article 38.

[4] Annex 9 defines a Recommended Practice as any specification, the observance of which has been recognized as generally practicable and as highly desirable to facilitate and improve some aspect of international air navigation, which has been adopted by the Council pursuant to Article 54 (*l*) of the Convention, and to which Contracting States will endeavour to conform in accordance with the Convention.

R. Abeyratne, *Regulation of Air Transport*, DOI 10.1007/978-3-319-01041-0_4,
© Springer International Publishing Switzerland 2014

necessary, international standards and recommended practices and procedures dealing with . . . customs and immigration procedures . . . and such other matters concerned with the safety, regularity and efficiency of air navigation as may from time to time appear appropriate". The policy with respect to the implementation by States of the Standards and Recommended Practices on Facilitation is strengthened by Article 22 of the Convention, which expresses the obligation accepted by each Contracting State "to adopt all practicable measures, through the issuance of special regulations or otherwise, to facilitate and expedite navigation by aircraft between the territories of Contracting States, and to prevent unnecessary delays to aircraft, crews, passengers, and cargo, especially in the administration of the laws relating to immigration, quarantine, customs and clearance", and by Article 23 of the Convention, which expresses the undertaking of each Contracting State "so far as it may find practicable, to establish customs and immigration procedures affecting international air navigation in accordance with the practices which may be established or recommended from time to time pursuant to this Convention".

The Standards and Recommended Practices on Facilitation inevitably take two forms: first a "negative" form, e.g. that States shall not impose more than certain maximum requirements in the way of paperwork, restrictions of freedom of movement, etc., and second a "positive" form, e.g. that States shall provide certain minimum facilities for passenger convenience, for traffic which is merely passing through, etc. Whenever a question arises under a "negative" provision, it is assumed that States will, wherever possible, relax their requirements below the maximum set forth in the Standards and Recommended Practices. Wherever there is a "positive" provision, it is assumed that States will, wherever possible, furnish more than the minimum set forth in the Standards and Recommended Practices.

The first Standard in the Annex in Chapter 2 broadly sets out the theme that Contracting States shall adopt appropriate measures for the clearance of aircraft arriving from or departing to another Contracting State and shall implement them in such a manner as to prevent unnecessary delays.[5] This provision in effect gives muscle to the meaning and purpose of consumer protection and assists airlines in establishing their reputation of on time departure and arrival. The provision is immediately supported by two Standards which say respectively that Contracting States shall not require any documents, other than those provided for in the chapter, for the entry and departure of aircraft and that Contracting States shall not require a visa nor shall any visa or other fee be collected in connection with the use of any documentation required for the entry or departure of aircraft.[6]

The Chicago Convention, in Article 13 provides that the laws and regulations of a Contracting State as to the admission to and departure from its territory of

---

[5] Standard 2.1. See also Standard 3.1 which provides that, in order to facilitate and expedite the clearance of persons entering or departing by air, Contracting States shall adopt border control regulations appropriate to the air transport environment and shall apply them in such a manner as to prevent unnecessary delays.

[6] Standards 2.5 and 2.6.

passengers, crew or cargo of aircraft, such as regulations relating to entry, clearance, immigration, passports, customs and quarantine shall be complied with by or on behalf of such passengers, crew or cargo upon entrance into or departure from, or while within the territory of that State. This provision ensures that a Contracting State has the right to prescribe its own internal laws with regard to passenger clearance and leaves room for a State to enact laws, rules and regulations to ensure the security of that State and its people at the airport. However, this absolute right is qualified so as to preclude unfettered and arbitrary power of a State, by Article 22 which makes each Contracting State agree to adopt all practicable measures, through the issuance of special regulations or otherwise, to facilitate and expedite navigation of aircraft between the countries.

4.1 Carriage of Persons

Everyone travelling by air across borders is aware that visa requirements are a severe hampering agent and that tourism in particular is seriously affected by stringent visa requirements imposed by States. More than one billion tourists crossed international borders during 2012, over half of who travelled by air to their destinations. The total number of international tourists, which includes both business and leisure travelers, is expected to reach 1.8 billion by 2030. Standard 3.17 provides that Contracting States shall not require exit visas from their own nationals wishing to tour abroad nor from visitors at the end of their stay. The Annex also does away with the tedious requirement that identity documents being presented in writing. It provides that Contracting States should not require either from visitors travelling by air, or from aircraft operators on their behalf, identification information in writing supplementary to that presented in their identity documents. Where the collection of identity information is required, Contracting States should develop systems for the electronic capture of this information from machine readable travel documents or other sources.[7]

Arguably, one valuable recommendation that was considered by ATConf/6 was that ICAO should undertake a cost–benefit analysis of measures aimed at enhancing connectivity, including those of facilitation. Such an analysis, it was claimed, would compare the economic cost of implementing those measures with the economic benefits that are likely to result from the implementation of those measures. Economic benefits would include obvious elements such as impact on gross domestic product (GDP) and job creation generated directly or indirectly (including induced and catalytic effects) by increased connectivity.

ATConf/6 was advised that, according to UNWTO and the World Travel and Tourism Council (WTTC), simplified visa facilitation processes in the G20 economies could lead to 102 million additional international arrivals and generate

---

[7] Recommended Practice 3.26.

US$206 billion in tourism receipts while creating as many as 5.1 million jobs in the G20 economies up to 2015. It was also advised that visa facilitation is central to stimulating economic growth and job creation through tourism. UNWTO and WTTC were of the view that the G20 can have a particularly important role to play in this respect. G20 economies could boost their international tourist numbers by an additional 122 million, generate an extra US$206 billion in tourism exports and create over five million additional jobs by 2015 by improving visa processes, according to preliminary research by UNWTO and the World Travel & Tourism Council (WTTC) presented on the occasion of the T20 Ministers' Meeting (Merida, Mexico, 16 May 2012). The Conference also noted that, as examples, the growth in air services between Poland and the United Kingdom (UK) since 2003 resulted in a gross domestic product (GDP) increase of 27 % for Poland, whereas the increase in the already well-served UK was a much smaller 0.5 %. These changes provide an estimated long-term boost to Poland's GDP of US$634 million per annum. The UK also benefited, with an estimated boost to its GDP of US$45 million per annum.

On 18 March 2013 ICAO and UNWTO signed a special joint statement on aviation and tourism in which the two Organizations pledged to optimize the benefits of aviation and tourism through: maximizing synergies between air transport and tourism, while finding ways to continually enhance collaborative endeavours; cooperating for the modernization of the air transport regulatory framework; and enhancing air transport connectivity further through cooperation. With regard to visa and other travel document formalities and issuance, including the simplification of visa processing and the development of multi-State regional visas and e-visas, ICAO and UN WTO pledged continued cooperation as well for the improvement of air passenger flow management at airports; and for the implementation of the Essential Service and Tourism Development Route (ESTDR) concept. They also vowed to cooperate in contributing to the emergence of globally convergent rules on the protection of passengers, tourists and tourism service providers, within their respective mandates and the framework of existing or future international bilateral or multilateral agreements. Another area of cooperation identified in the joint statement was in contributing to the reduction of greenhouse gas emissions from aviation and tourism and giving due consideration to the particular importance of air transport for tourism development in long-haul destinations and landlocked or island countries as well as assessing the impact of taxes, charges and other levies on aviation and tourism, and thus on global economic growth.

Undoubtedly, these are lofty and necessary goals. But goals alone are not enough. There has to be an action plan with Strategic thrusts that each party could forge. The first step would be to conduct a study that brings out a cost–benefit analysis of the benefit of facilitation on air transport, and, contrary to the recommendation of ATConf/6, this should not be done by ICAO alone but jointly— between ICAO, UNWTO and IATA.

ICAO has initiated and developed many tools of facilitation and security that go towards ensuring efficient and economical air transport. Noteworthy among these are the machine readable travel document (passport and visa)[8] and the e-Passport and the Public Key Directory.[9] The ICAO Facilitation Manual defines the ePassport as a machine readable passport that has a contactless integrated circuit embedded in it and the capability of being used for biometric identification of the machine readable passport holder in accordance with the Standards specified in the relevant part of ICAO document 9303 (Machine Readable Travel Documents).[10] ePassports are easily recognised by the international ePassport symbol on the front cover.[11] The key consideration of an ePassport is Global Interoperability, which in turn would ensure that the use of the ePassport would ensure seamless air transport across borders, thus making air transport a more efficient product than what it is now. These tools would make sure for States and other key stakeholders of air transport that the time required for the accomplishment of border control procedures in respect of persons and freight (the release and clearance of goods) will be kept to a minimum; persons travelling and freight forwarders and shippers sending their goods by air would be experiencing minimum inconvenience caused by bureaucracy; there will be greater exchange of information between those involved in transporting persons and goods; and optimum levels of security and expedience will be assured.

Annex 9, in Standard 3.7 requires ICAO member States to regularly update security features in new versions of their travel documents, to guard against their misuse and to facilitate detection of cases where such documents have been unlawfully altered, replicated or issued. Recommended Practice 3.9 suggests that member States incorporate biometric data in their machine readable passports, visas and other official travel documents, using one or more optional data storage technologies to supplement the machine readable zone, as specified in Doc 9303, Machine Readable Travel Documents. The required data stored on the integrated circuit chip is the same as that printed on the data page, that is, the data contained in

---

[8] See Abeyratne (1992), pp. 1–31.

[9] Abeyratne (2005), pp. 255–268. Over 104 States are currently producing and using ePassports and there are approximately 400 million in circulation. This accounts for 33 % of all passports used globally. The additional feature that the ePassport carries in the conventional machine readable passport is a chip containing biometric and biographic information which have to be validated accurately, efficiently and quickly while retaining the security and integrity of the information.

[10] See The Facilitation Manual, Doc 9957, ICAO: Montreal, First Edition 2011, Definitions at X. ICAO has been working on the development of passports since 1968. The Seventh Session of the ICAO Facilitation Division in 1968 recommended that a small panel of qualified experts including representatives of the passports and/or other border control authorities, be established: to determine the establishment of an appropriate document such as a passport card, a normal passport or an identity document with electronically or mechanically readable inscriptions that meet the requirements of document control; the best type of procedures, systems (electronic or mechanical) and equipment for use with the above documents that are within the resources and ability of Member States; the feasibility of standardizing the requisite control information and methods of providing this information through automated processes, provided that these processes would meet the requirements of security, speed of handling and economy of operation.

[11] http://www.dhs.gov/xtrvlsec/programs/content_multi_image_0021.shtm.

the machine-readable zone plus the digitized photographic image. Fingerprint image(s) and/or iris image(s) are optional biometrics for member States wishing to supplement the facial image with another biometric in the passport. Member States incorporating biometric data in their Machine Readable Passports are to store the data in a contactless integrated circuit chip complying with ISO/IEC 14443 and programmed according to the Logical Data Structure as specified by ICAO.

4.2   Carriage of Cargo

In the carriage of cargo, as in the carriage of persons addressed in Standard 2.1 of Annex 9, the aim is to facilitate and expedite the release and clearance of goods carried by air, and for Contracting States to adopt regulations and procedures appropriate to air cargo operations and to apply them in such a manner as to prevent unnecessary delays.[12] The ICAO/IATA/ACI cost–benefit analysis on savings accrued by cutting delays in the carriage of 48 million tons of cargo that were carried in 2011 would hopefully reveal impressive figures of efficiency in the carriage by air of cargo. Standard 4.11 of Annex 9 prescribes that, subject to the technological capabilities of the Contracting State, documents for the importation or exportation of goods, including the Cargo Manifest and/or air waybills, shall be accepted when presented in electronic form transmitted to an information system of the public authorities. To facilitate electronic data interchange, Contracting States are required to encourage all parties concerned, whether public or private, to implement compatible systems and to use the appropriate internationally accepted standards and protocols.[13]

The World Customs Organization (WCO) is promoting its Framework of Standards to Secure and Facilitate Global Trade (SAFE) and many of its member states have signed letters of intent to implement it. As a result more countries will be requiring advance cargo information. This would require measures that are calculated to ensure avoidance of costly data re-entry tasks; promotion of a paperless environment; improvement of existing industry processes by enhancing the Electronic Data Interchange (EDI) environment; recognizing other governments upcoming Customs regulations; and avoidance of a wide variability of requirements. For this a standard approach to information requirements is necessary.

In its World Air Cargo Forecast for 2012–2013 Boeing suggests that over the next 20 years, world air cargo traffic will grow 5.2 % per year and that air freight, including express traffic, will average 5.3 % annual growth, measured in RTKs.

---

[12] Annex 9, Standard 4.1.

[13] *Id.* Standard 4.15. Recommended Practice 4.16 adds that electronic information systems for the release and clearance of goods should cover their transfer between air and other modes of transport.

Air mail traffic will grow much more slowly, averaging only 0.9 % annual growth through 2031. The overall forecast for air cargo traffic is that it will increase from 202.4 billion RTKs in 2011 (down from its 2010 record of 204.2 billion RTKs) to more than 558.3 billion RTKs in 2031.

Boeing also forecasts that Asia will continue to lead the world air cargo industry in average annual growth rates, with domestic China and intra-Asia markets expanding 8.0 and 6.9 % per year, respectively. Latin America markets with North America and with Europe will grow at approximately the world average growth rate, as will Middle East markets with Europe. The more mature North America and Europe markets reflect slower and thus lower-than-average traffic growth rates. As for Europe, approximately 14.5 million tonnes of air freight (both national and international) was carried through airports within the EU-27 in 2011— Airports in Germany handled 4.3 million tonnes of air freight, considerably more than in any other EU Member State—the United Kingdom had the second highest amount of air freight at 2.4 million tonnes. Some of the smaller EU Member States are relatively specialized in air freight, notably all of the Benelux countries, and in particular, Luxembourg (which ranked as the seventh largest air freight transporter among the EU Member States).[14] These figures present all the more reason that a comprehensive and authentic study has to be undertaken that would reflect the tangible benefits of facilitation to air transport.

References

Abeyratne RIR (1992) The development of the machine readable passport and visa and the legal rights of the data subject. Ann Air Space Law XVII(Part II):1–31

Abeyratne RIR (2005) The e-passport and the public key directory – consequences for ICAO. Air Space Law XXX(4–5):255–268

---

[14] See http://epp.eurostat.ec.europa.eu/statistics_explained/index.php/Freight_transport_statistics#Air_freight.

Chapter 5
Consumer Rights

The Chicago Convention refers to consumer rights and not to consumer protection,[1] when in Article 44 (d) it mentions the needs of the people of the world for air transport. Basically what consumer rights means in this context is that the travelling public needs availability of service, at a good price, which renders value for money. Neither ATConf/5 nor ATConf/6 addressed this glaring issue either directly, or indirectly. This is indeed unfortunate. Instead, both conferences addressed consumer protection, which involved such issues as the availability of flight information at the airport, overbooking, delayed transport et al.

5.1 Consumer Protection

ATConf/5 in 2003 addressed the subject of consumer rights under the guise of consumer protection, but only just. The Conference noted that various States had different approaches to the rights of the consumer. Those States which regarded air transport as a public utility emphasized heavily on regulation whereas States which considered air transport as a predominantly commercial activity relied on competition. Low cost carriers were considered as offering incentives of low fares to the consumer which in turn was perceived by some as a measure of consumer protection. European rules were recognized by some participants as offering equitable regulations supporting and encouraging consumer protection.

The Conference concluded that States should minimize differences in the content and application of regulations, with a view to avoiding legal uncertainty that could arise from the extra-territorial application of national laws. As recommended by ATConf/5, and in order to assist States, the Secretariat developed a summary of airline and governmental responses to some of the more prominent consumer protection issues. At its 37th Session (Montreal, 27 September to 8 October

---

[1] For a comprehensive discussion on consumer protection see Abeyratne (2012), at pp. 103–146.

R. Abeyratne, *Regulation of Air Transport*, DOI 10.1007/978-3-319-01041-0_5,
© Springer International Publishing Switzerland 2014

2010) the ICAO Assembly adopted Resolution A37-20 on ICAO's continuing policy in the air transport field, which includes notably a new clause stating that "consumer interests should be given due regard in the development of policy on regulation of international air transport."

At ATConf/6 in 2013, the Conference recognized that regulatory activity has expanded rights for issues linked to events such as flight delays, cancellations, and overbooking. These rules sometimes coexist with voluntary airline commitments, and have been applied in cases of massive travel disruptions. In 2004, the European Union (EU) adopted Regulation No. 261/2004, *Establishing common rules on compensation and assistance to passengers in the event of denied boarding, cancellation or long delay of flights*, the scope of which has been broadly construed in decisions by the European Court of Justice. In 2011, the European Commission decided to launch a public consultation on the possible revision of Regulation No. 261/2004. The United States (U.S.) Department of Transportation (DOT) has promulgated a 2011 rule, applying to U.S. and foreign air carriers operating aircraft of 30 seats or more to or from a U.S. airport. The rule increases compensation for passengers involuntarily denied boarding and establishes a maximum time (i.e. 3 h for domestic flights and 4 h for international flights, during which time an aircraft may stay on the tarmac without allowing passengers to disembark).

Various consumer protection initiatives have been taken in other regions. In November 2004, the Latin American Civil Aviation Commission (LACAC) adopted Recommendation A16-8 (*Rights of the Users*), urging its Member States to promulgate a set of rules protecting passengers holding confirmed reservations who have been denied boarding involuntarily. The regime provides passengers with the option to choose between reimbursement and alternative transport (including assistance relating to food/drinks, communications, and accommodations). China (Rules of civil aviation passenger and baggage, Decrees No 49 and 70 CAAC), Saudi Arabia (Consumer Protection Regulation in 2005) and Israel (Airline Passenger Rights law in 2012) have also developed consumer protection rules. In Singapore, the Civil Aviation Authority has taken a different approach by working with consumer representatives to educate passengers on the key aspects of air travel, including what to consider when purchasing airline tickets and recourse options in the event of an airline service lapse.

The World Tourism Organization (UNWTO) is currently developing a draft Convention on the Protection of Tourists and Tourism Service Providers. The proposed scope of the draft instrument is concerned with, inter alia, State assistance obligations in "force majeure" situations, the protection of the tourist in the event of insolvency of the travel organizer, as well as package travel related issues. It should be noted that a definition for the term "force majeure", as used in the draft instrument, is also under development. Bearing in mind that more than 50 % of tourists are air travellers, the ICAO Secretariat is collaborating with the UNWTO with a view to avoiding duplication of efforts or inconsistencies with existing air law instruments. The multiplicity of "unbundled" service options has complicated price transparency, which in turn has triggered substantial regulatory activity.

ICAO exhorted ATConf/6 to agree that ICAO develop a set of core principles on consumer protection under a structured framework and continue to play a leadership role in developing policy guidance to address emerging issues concerning airline consumer protection at the global level, taking into account the interests of States, the industry, air travellers and other aviation stakeholders. Another task ICAO took upon itself was to monitor consumer protection issues with a view to enabling States to decide, in due course, on the need for the development of a more formal arrangement, such as that of a "global code of conduct" and continue to cooperate and coordinate with other international organizations concerned in order to avoid an overlap of efforts, particularly with respect to UNWTO work presently underway.

UNWTO advised the Conference that according to the decision of the World Tourism Organization (UNWTO) Executive Council at its 89th session (Kish Island, Iran, 24–26 October 2010) the "*UNWTO working group on the protection of tourists/consumers and travel organizers*" was set up in April 2011. This was a response to the insufficiency of existing binding rules at a global level governing the rights and obligations of tourists/consumers and of travel organizers, a fact so clearly brought to light in April 2010 by the crisis provoked by the eruption of the Icelandic volcano and its tremendous effect on the travel and tourism sector. The working group consists of official expert delegates from Member States of every region, representation of the tourism private sector such as the International Air Transport Association (IATA), the World Association for Professional Tourism Training (AMFORT), the International Hotel & Restaurant Association (IH&RA), the United Federation of Travel Agents' Associations (UFTAA), the World Travel Agents Association Alliance (WTAAA), the Group of National Travel Agents' and Tour Operators' Associations within the EU (ECTAA), the European Guarantee Funds' Association for Travel and Tourism (EGFATT), the Hotels, Restaurants & Cafés in Europe (HOTREC), the World Travel & Tourism Council (WTTC), the International Automobile Federation (FIA), the Airports Council International (ACI), international organizations such as ICAO, and representatives of the European Commission. In line with the decision of the UNWTO General Assembly (Gyeongju, Republic of Korea, 8–14 October 2011), the working group is in the process of elaborating an international convention on the protection of tourists and tourism service providers.

The resulting Draft Convention will have a general part, which introduces the general principles, the scope elements, the main definitions and the necessary international law related provisions (amendment rules, entry into force, reservations, signature, ratification rules, deposit, etc.). The Annexes of the convention will include standards as binding rules and recommended practices as non-binding rules. Annex I of the draft convention contains the assistance obligations of States Parties in *force majeure* situations indicating clear distinction from private sector's obligations. Annex II deals with package travel issues in detail. It includes *force majeure* situations, minimum liability rules for non-performance and improper performance, and minimum information which should be provided to the tourists before the conclusion of the package travel contract and the minimum content elements of the contract as well. Finally, in case of the insolvency of the service providers minimum rules are established for the protection of tourists (financial security). With the consent of the related stakeholders' organizations, Annex III on accommodation related provisions focuses on mainly information obligations in accordance with the existing regional and global regulatory frameworks.

UNWTO stressed that there was a need for further strengthening the cooperation between ICAO and UNWTO on tourists/consumers protection activities. It suggested strongly that the close working relationship should be maintained via the UNWTO working group's activities, bilateral negotiations, etc. in order avoid inconsistencies and possible duplication of efforts, as well as to minimize conflict in related rules and regulations. UNWTO also supported the development of a set of core principles dealing with air passengers' rights and protection at a global level under the form of the indicative framework proposed in Appendix B of its paper.[2] UNWTO agrees that price transparency issues should be taken into consideration by the elaboration of these international guidelines.

The United States provided ATConf/6 with an array of measures that it had adopted on consumer protection in accordance with its law which prohibits unfair or deceptive practices or unfair methods of competition in air transportation or the sale of air transportation.[3] The law prohibits discrimination in airline service on the basis of disability. The United States Department of Transportation (DOT) adopts and enforces consumer protection and civil rights regulations, which are found in Title 14 of the Code of Federal Regulations (CFR). Accordingly, The United States proposed that States agree to certain customer service commitments beginning with adopting consumer protection regulations; and cooperating to seek compatibility in their respective consumer protection regulations. ICAO was requested to collect and disseminate the airline passenger consumer protection laws and regulations of States; and with the assistance of experts from interested States and stakeholders, develop a set of general core principles on airline passenger consumer protection.[4]

The African Civil Aviation Commission (AFCAC) advised (on behalf of its 54 member States) the Conference that consumers should be protected from airlines who become insolvent and who could not offer air transport to passengers and shippers who had paid for such services. They also said that measures introduced in various jurisdictions to protect the consumer include the elimination of misleading advertisements and the requirement to disclose all the facts with regard to travel a passenger is pays for. Furthermore AFCAC pointed out that the practice of free seating, which is a prolific practice in Africa has been questioned in draft regulations being developed.[5]

---

[2] A Tourism Perspective on Consumer Protection, *ATConf/6-31*.

[3] 49 U.S.C. 41712.

[4] Achieving Compatibility in Consumer Protection Regulations, *ATConf/6-WP/45*.

[5] African Air transport and Protection of the Consumer, *ATConf/6-WP/47*.

The European Union, which has adopted clear rules on consumer protection laid out before ATConf/6 its criteria for consumer protection. They were: non-discrimination in access to air transport which should include preventing air carriers from discriminating during ticketing by nationality, residence or disability. The EU said further that for persons with disabilities the provision of access and assistance has to be granted without any additional charges. Such persons should not be refused carriage on board an aircraft except on reasons of safety, which air carriers would have to justify. The EU also insists on transparency which should include the right for the passengers to have accurate, timely and accessible information in particular to: be able to obtain information on what is included within the price of their ticket. Passengers should have access to all relevant information (e.g. on price and charges applicable) in order to make a reasoned decision before buying an air ticket. Such information will enable them to fairly compare ticket prices and provides a level competitive playing field for industry. This approach would also ensure that air carriers detail the conditions and restrictions on any fare offered and those tickets are sold inclusive of all taxes, fees and charges, together with any "optional" charges—such as the carriage of luggage. This would prevent air carriers from advertising "tax-free" prices where charges that passengers have to pay are only added during the process of payment. Consumers should be advised before departure on which air carrier they will be flying and receive appropriate information before the purchase of their flight ticket and at appropriate stages of travel, particularly when disruption occurs. They should, under EU practice renounce travelling and obtain a full refund when the trip is not undertaken as planned due to the actions of the air carrier.

The EU also brought to the attention of the Conference the need to ensure immediate and proportionate compensation and assistance. This includes compensation to be offered to the passengers under special circumstances, such as denial of boarding, short-notice cancellation of flights and under certain conditions long delays. It also includes the right for passengers to have assistance at departure or at connecting points in the form of, for example: right to care, especially meals, refreshments, telephone calls, hotel accommodation and transport between the airport and place of accommodation; and the right to rerouting or reimbursement and rebooking.[6]

There is some jurisprudence elaborating consumer protection both in Europe and in the United States. The Grand Chamber of the Court of Justice of the EU (CJEU), in joint cases decided in October 2012[7] confirmed EU Regulation 261/2004 should be interpreted as entitling passengers to compensation if, due to a delay on the part of the airline, they arrive at their destination 3 h or more later than scheduled. The carrier could exonerate itself only if it proved that the delay was caused by extraordinary circumstances beyond the carrier's control which could not have been avoided even if the carrier took all necessary measures to avoid the delay.

---

[6] Basic Principles for Consumer Protection, *ATConf/6-WP/55*.

[7] *Nelson* v. *Lufthansa*, C581/10, and *British Airways, Easyjet and IATA* v. *UK CAA*, c629/10 TUI.

The court attributed its arrogation of powers in handing down its decision to what it called the principle of equal treatment. The Court aligned itself to Article 19 of the Montreal Convention of 1999 which stipulates that the carrier is liable for damage occasioned by delay in the carriage by air of passengers, baggage or cargo. The provision has a disclaimer, that the carrier shall not be liable for damage occasioned by delay if it proves that it and its servants and agents took all measures that could reasonably be required to avoid the damage or that it was impossible for it or them to take such measures.[8]

The CJEU in these decisions followed the 2009 *Sturgeon* judgment where the court held that although Regulation 261 did not expressly confer any right to fixed-rate compensation to passengers whose flights are delayed, it should be interpreted consistently with the EU law principle of equal treatment, and in this instance held that passengers whose flights are delayed for long periods are inconvenienced in the same manner as those whose flights are cancelled and therefore should be accorded the same treatment.

The joint cases and their decisions followed two other decisions handed down in the same month by the CJEU on denied boarding. In *Lassooy* v. *Finnair*[9] due to a strike by the staff at Barcelona airport, Finnair was forced to cancel its flight from Barcelona to Helsinki on 28 July 2006. Finnair decided to reschedule other flights to ensure that the passengers on the questioned flight were not too delayed. As a result, Mr. Lassooy was denied boarding when he presented himself for the flight, due to the fact that Finnair had accommodated passengers from earlier flights who had been "knocked on". CJEU held that "the applicable EU regulation extended beyond overbooking in cases of delayed transportation to operational factors and that Finnair was precluded from invoking the extraordinary circumstances" exception. In *Rodriguez* v. *Iberia*[10] which concerned an Iberia flight from Corunna (Spain) to Santo Domingo (on the route Corunna–Madrid, Madrid–Santo Domingo comprising a trip of two flights) on which Mr. Rodriguez and another passenger were booked to travel, after the passengers had been given their two boarding cards each by the Iberia check in counter, the latter card was cancelled by Iberia owing to a delay (of 1 h and 25 min) on the first leg of the flight. Although the two passengers presented their cards at the counter at Madrid, they were not allowed to board on the ground that their boarding cards had been cancelled and their seats allocated to other passengers. As a result Mr. Rodriguez and the other passenger reached their destination 27 h after the original scheduled time of arrival in Santo Domingo. CJEU held that the cause of the denied boarding was the carrier's decision to cancel the boarding cards for the connecting flight and its mistaken belief that the two passengers may not make it to the connecting flight. CJEU held that the concept of

---

[8] Convention for the Unification of Certain Rules for International Carriage by Air, signed at Montreal on 28 May 1999. *ICAO Doc. 9740*.

[9] See *Legal Eye – An Overview of Recent Developments*, Stephenson Harwood: Autumn/Winter 2012 at 4.

[10] *Ibid.*

denied boarding as reflected in Regulation 261 was not restricted to overbooking but extended to other factors as well including operational reasons. In this case, Iberia had no reasonable grounds to exonerate itself from liability.

Across the pond, there are United States decisions which adopt the same trend. In the 2010 case of *Azza Eid et al.*; v. *Alaska Airlines*[11] the court held that airlines who force passengers to disembark could e held liable for damage caused. In the same year, the US Department of Transport (DoT) through its Aviation Consumer Protection and Enforcement Division, which is tasked with enforcing consumer protection laws against domestic and foreign carriers operating into the United States, issued 27 orders against air carriers for tarmac delays, to pay fines amounting to $1.7 million. In 2011, the figure went up to 47 orders amounting to $3.26 million in penalties. The passenger's fault or contributory negligence often vitiates the carrier's liability as was seen in the September 2012 decision in *Giuffre* v. *Delta Airlines*[12] where the U.S. District Court for the Eastern District of New York held that failure to check in on time was fatal to passenger's claims. In this case the passengers had not been able to fill in the necessary forms 1 h prior to boarding, due to their late arrival at the check in counter.

Another dimension to consumer protection is pricing. In Europe, there is entrenched legislation of the European Union concerning predatory pricing, competition and fair trade. Article 82 of the *Treaty of Rome* prohibits abuse of dominant position by a carrier through a determination of the relevant market and market share enjoyed by the carrier under evaluation. This information is matched with the pricing practices of the carrier which must not be lower than average variable cost. The Competition Act of 1998 of the United Kingdom links predatory pricing with dominant position and uses a process similar to that of the European Union[13] in assessing price–cost relationships. Germany has similar legislation in the *Gesetz gegen Wettbewerbsbeschränkungen* (GWB) which is the Act Against the Restraint of Competition, which identifies predatory practices as an abuse of dominant position if the predator is dominant in the market; the conduct of predatory pricing is sustained and continuous and pricing is below average costs without objective justification.[14]

The United States competition law as its genesis the *Sherman Act* of 1890 followed by the *Clayton Act* of 1914 (which was later amended in 1936). Such established legislation has been interpreted judicially to require two criteria: pricing must be below average variable costs and there has to be proof of recoupment of losses incurred during the alleged period of predatory pricing. In the 2001 case of

---

[11] No. 06-16457, D.C. No CV-04-01304-RCJ.

[12] No 10-cv-1462, slip.op (E.D.N.Y. September 2012).

[13] Both the EU and the United Kingdom uses the AKSO NV case as a benchmark where a Dutch chemical company, with a 65 % market share of its flour bleach product was found to be abusing its dominant position. The European Court of Justice found that price below average variable cost by means of which dominant competitor seeks to eliminate its competition is regarded as an abusive practice. See *AKZO Chemie BV* v. *EC* (1991) ECR 1-3359 at paras. 71–72.

[14] *Gesetz gegen Wettbewerbsbeschränkungen* (GWB) Section 20(4).

US v. *AMR Corp*[15] the court held that an air carrier matches prices and increases output when faced with competition from low cost carriers is not guilty of monopolization of the market even if the carrier reverted to its original level of pricing after the low cost carrier concerned had left the market. The court based its decision on the basis that the carrier had not price its fare at a level below a appropriate level of cost. The carrier, in this instance, was found to be meeting the competition fairly rather than undercutting the other carrier, and there was no evidence that the carrier would recoup its losses through supra competitive pricing. Predation in Canada is brought within the purviews of both civil and criminal law where, Section 50(1) (c) of the Canadian Competition Act recognizes selling at an unreasonably low price an act of predation when such practice is calculated to eliminate competition or lessen a competitor's ability to compete.

The Australian *Trade Practices Act* of 1974, which is administered through the Australian Competition and Consumer Commission, provides in Section 46 that, when a firm takes control of dominant market power, particularly with intent to lessen or eliminate competition, the onus is on the person holding the position of dominance to prove his actions are not tantamount to predatory practices. The criterion used is that recoupment through pricing at supra competitive levels was a *sine qua non* to prove predatory pricing.

The above discussion leaves no room for doubt that there is strong regulatory control of fares and services offered to the consumer in air transport. The responsibility in this regard lies primarily in sovereign States who act through ICAO for global consensus and also by themselves to enact national legislation and adopt national policy.[16] The above notwithstanding, the most compelling area for consideration, particularly from a legal perspective might be the services provided, particularly by a low cost carrier who does not provide the usual frills of international air transport. In a point to point service, usually obtained through the internet, where support services at the airport would be minimal and complaints from the passengers or consignors/consignees would be difficult to channel to the carrier, it would indeed be interesting to envision how courts would handle a particular case of passenger injury or the action brought before them for litigation.

5.2 Consumer Rights

As already stated, the Chicago Convention is about consumer rights. No one would deny that air travel is price sensitive and discretionary. Airlines have to run a profitable business amidst rising fuel costs, which have more than tripled over the past 5 years. This has made the operating cost structures of airlines flip from fixed to variable costs. All this to say that airlines are becoming more and more strategic in

---

[15] U.S. District Court, District of Kansas, 27 April 2001, 28 Avi 15,204.

[16] Shaw (2003) at pp. 694–697.

ensuring their profit margins, often at the expense of the consumer. As a result, connectivity, the choice of flights covering the shortest distance to the destination at the correct price yielding value for money are often denied to the traveller. The earlier discussion on revisiting Article 6 of the Chicago Convention and the chapter on connectivity elaborated this point.

Airlines have to be competitive and liberal, and their States of nationality have to recognize the overall need of serving the public over pandering to the survival of their airlines by preventing competition among carriers. An example is Singapore which has concluded air services agreements with more than 100 countries and territories, including about 40 open skies agreements. Singapore adopts a liberal aviation policy to allow airlines full flexibility to respond to market opportunities. This enables passengers and shippers to have the widest possible travel and flight options at competitive rates so that Singapore continues to be the transfer hub of choice for international travellers and shippers. Open skies agreements allow carriers to operate any number of flights between and beyond both signatory states, enabling them to tap traffic from third countries to improve the commercial viability of scheduled flights. Notably, in November 2007, Singapore concluded a landmark Agreement with the United Kingdom that includes unlimited "hubbing" and cabotage rights.

Another example is Qatar whose mission is pro-aviation government policy, industry–government partnership and a vision that embraces the changing industry dynamics driven by globalisation. Emirates Airline is run as a fully commercial business and is treated like any other airline operating into Dubai International in terms of airport and landing charges. Paul Griffiths, CEO of Dubai Airports said:

> Industry partners must learn to cooperate and coordinate activities to better bring their individual strengths to the table. Governments must adopt policies that support liberalisation and sustainable growth. And both must commit to developing a lasting partnership that recognises the changing face of our industry and seeks the greatest efficiencies from an evolving global network.[17]

Abu Dhabi-based airline Etihad Airways of the United Arab Emirates has applied for an open skies agreement with Canada. Etihad is keen to be able to expand its services out of Toronto. Chief Executive James Hogan said at an interview: "[In the U.A.E.] we have an open skies environment and we would love to see a Canadian airline fly anywhere within the U.A.E. every day".[18]

By 2010, the UAE, Dubai and Emirates Airline had secured over 60 open or highly liberal aviation agreements, following further liberalization success in Latin America, Africa and Europe. According to According to Tim Clarke, President of Emirates:

> Open skies secured by the UAE and Dubai Civil Aviation Authorities with key economies now represent the majority of Emirates Airline's air services access worldwide, with an average of six new open or liberalized deals now being signed a year... The world, often

---

[17] http://www.dubaiairport.com/en/media-centre/pages/press-releases.aspx?id=10.

[18] http://www.eturbonews.com/5153/abu-dhabi-airline-seeks-open-skies-canada.

led by emerging markets, is liberalizing faster than many believed possible. This is good news for consumers, traders, exporters and travelers generally.[19]

When Emirates commenced its operations to Australia in 1997, the airline was viewed with trepidation and concern by QANTAS, as a threat to its market share. This concern was shared by the Australian authorities. However, attitudes quickly changed, and this concern was obviated when they realized the added economic benefit quickly enjoyed by the places Emirates flew to. Currently, Emirates operates 49 flights a week to Australian cities and hopes to expand this number to 80. QANTAS and Emirates are now partners and that. . .is the way to go.

Emirates was the most profitable airline in 2009/2010 earning a net profit of $963 million during the 2 years earned by carrying 27.5 million passengers in 124 wide body aircraft. In the Emirates 2010–2011 financial report, it was reported that with 182,757 million available seat kilometres, the airline was the world's largest airline by scheduled international passenger kilometres flown. There is no room for doubt that Emirates' success is due first to the aviation policy it follows, through visionary leadership, competent management and, above all, open competition that offers ready availability of service through connectivity; the right price; and value for money.

In my earlier publications I have ask the fundamental question—as to whether governance in aviation has been for the past 68 years on the right track and if it has not, how we can get it back on track. I am here not referring to corporate governance, or to ICAO, but to performance governance, otherwise known as business governance, which speaks to performance and value creation through resource utilization.

What do we stand for as a global aviation community? How do we strike the balance between growth and development? Where does aviation and its governance fit into a world transformed by the winds of globalization and change through technology?

Governance, as we know, is a set of responsibilities and practices that are aimed at achieving strategic direction and ensuring that objectives are achieved. The question then is: "have we a strategic direction and have we achieved our goals in aviation?"

Let us start at the beginning. What is aviation's strategic direction? As far back as 1944, the world, represented then by 52 signatory States to a multilateral treaty called the Chicago Convention—which now has 191 States acceding to it—called aviation's strategic direction "creating and preserving friendship and understanding among the nations and people of the world". One would not imagine, for a moment, that there is written anywhere in a global document that aviation's strategic direction is to make as much money as possible to the exclusion of others or to give priority to the interests of States or the air transport industry. Creating and preserving friendship and understanding among people can only be achieved through optimum connectivity.

---

[19] http://www.mb.com.ph/articles/292885/uae-dubai-emirates-airlines-secure-60-open-skies-agreements.

Having given this direction, the same multilateral treaty goes on to say that no scheduled international air service may be operated over or into the territory of a contracting State except with the special permission or other authorization of that State. It is a curious fact that in 1609, Hugo Grotius wrote in his magnum opus *Maré Liberum* (free seas) that the oceans should be open to sea faring by anyone. Yet more than three centuries later, precisely the opposite principle was adopted in 1944 by the aviation powers that assembled in Chicago.

The trouble with air transport is that, while on the one hand it is a product, on the other hand regulations pertaining to this product may constrain its availability to the consumer by depriving him of the various choices of air travel he might have under a liberalized system. In other words, State policy and the protection of national interests take precedence over the interest of the user of air transport. The aviation industry offers only one product to the ultimate consumer and that is the air transport product.

We know the air transport industry is cyclical and is profoundly affected by the world's economic health. I need not elaborate the economic vicissitudes of Europe and the significant growth elsewhere in Asia.

Aviation is a global industry and the need for air transportation continues to grow, as major cities continue to grow in population and prosper. A recent forecast has revealed that while in the 1970s there were just four major agglomerations of over ten million people there are 26 today and there will be more than 30 by 2015. As economic prosperity grows more and more people will demand access to air transport and traffic growth is expected to double in 15 years.

"Connectivity" which is the most compelling need in aviation, and embodied in the Chicago Convention as *inter alia* "meeting the needs of the people of the world for efficient and economical air transport" is stultified by interests of commercial and national policy. Let me give just one example where this insurmountable obstacle has been overcome:

When Emirates commenced its operations to Australia in 1997, the airline was viewed with trepidation and concern by QANTAS, as a threat to its market share. This concern was shared by the Australian authorities. However, attitudes quickly changed, and this concern was obviated when they realized the added economic benefit quickly enjoyed by the places Emirates flew to. Currently, Emirates operates 49 flights a week to Australian cities and hopes to expand this number to 80. QANTAS and Emirates are now partners and that. . .is the way to go.

I might also add that the air services agreement between the UAE and the United States allows Emirates to operate to any point in the States, how often they wish with no capacity restriction, and with rights to carry traffic from intermediate points.

At the Chicago Conference in 1944, the United States proposed a multilateral agreement calculated to guarantee commercial landing rights everywhere in the globe to all the world's airlines without restriction. The United States took the position that the use of the air and the use of the sea were both common in that they were highways given by nature to all men. They were different in that man's use of the air is subject to the sovereignty of nations over which such use is made. The United States was therefore of the opinion that nations ought to arrange among themselves for its use in such manner as would be of the greatest benefit to all humanity, wherever situated

The United Kingdom said:

> While recognizing national interests we want to encourage enterprise and efficiency which are indeed themselves a national as well as an international interest. And we want therefore to encourage the efficient and to stimulate the less efficient. . .only by common action on some such lines as indicated can we reduce and gradually eliminate subsidies, thereby putting civil aviation on an economic footing and incidentally very considerably relieving the tax payer. Unrestricted competition is their most fruitful soil.

The United Kingdom seems to have adopted a balanced approach that supported the establishment of air services to serve the needs of the travelling public, while not unduly affecting the rights of States to have a fair share of traffic for themselves.

India said:

> We believe that the grant of commercial rights - that is to say, the right to carry traffic to and from another country, - is best negotiated and agreed to on a universal reciprocal basis, rather than by bilateral agreements. We think that only such an arrangement will secure to all countries the reciprocal rights which their interests require.

Here's what Canada said at the Conference:

> We (Canada) believe that there must be greater freedom for development of international air transport and that this freedom may best be obtained within a framework which provides equality of opportunity and rewards for efficiency.

Let us take this discussion to a different dimension—and that is economics. In June 2012, at the IATA Annual General Meeting in Beijing, it was revealed that in 2011, aviation safely transported some 2.8 billion passengers and 48 million tons of cargo. It was also noted that the value of goods transported by air was estimated at $5.3 trillion, which equals to 35 % of the value of all goods traded internationally. A recent study by Oxford Economics has confirmed that aviation's contribution to the global economy supports 57 million jobs and some $2.2 trillion in economic activity. Oxford Economics projected that aviation will grow about 5 % annually to 2030. That would see passenger numbers rise to 5.9 billion and cargo shipments could triple to nearly 150 million tonnes. This connectivity would support 82 million jobs and $6.9 trillion of global GDP. If growth is held back by even one percentage point, the global economy would forfeit 14 million jobs and over $1 trillion in GDP contribution from aviation. Aviation's benefits are not guaranteed. Aviation is expected to grow about 5 % annually to 2030. If that growth is held back by even one percentage point, the global economy will forfeit over a trillion dollars and 14 million jobs.

The critical statement here is "If that growth is held back by even one percentage point, the global economy will forfeit over a trillion dollars and 14 million jobs". The other side of the coin is that improved connectivity—in other words liberalization of air transport as against protectionism—promotes growth, instead of holding it back.

Two good examples are Dubai and Singapore. Both these are small city States with small local populations. But both their airports see more than 30 million passengers pass through their gates every year. Both countries practice an open skies policy with no demand of reciprocity, which means that foreign airlines have free entry. This is what one would call connecting the world and meeting the needs of the people of the world.

Air transport does not have a smooth trading passage. In 2009 IATA summed up the predicament of the air transport industry succinctly when it says of the past.

A hyper-fragmented global industry was struggling for its survival. National flags were originally put on the tail to protect the airlines. That was an age of regulated travel for the rich. Today we are an industry that moves over 40 million tonnes of cargo and 2.2 billion people annually. Our activities support 32 million jobs and US$3.5 trillion in economic activity. The rules set to protect this industry do not work in today's environment. Even in good economic times, the industry has not covered its cost of capital. Restrictions on ownership and market access have prevented airlines from growing into strong global businesses.

The latest crisis is yet another reminder that there is no policy purpose in keeping the industry financially weak with "out-dated restrictions".

One could well argue that the trouble with air transport has been that it has always been about State policy and interests of States and not about the rights of the consumer. Restrictions on Foreign direct investment (FDI) through rigid ownership and control policy further strengthen this approach. FDI promotes economic growth and facilitates competition.

Caps on FDI serve to obviate the need for governments to invest, particularly when they have to protect ownership and control of nationals in designated airlines. This is counter-intuitive as there will not be ownership and control for governments to protect if the airlines go bankrupt for want of capital. To add to the problem, the credit crunch of 2008 and rising fuel prices made the availability of capital even scarcer.

Generally, perhaps with the exception of air transport within Europe, flag carriers are effectively protected by a regulatory process driven by a network of bilateral agreements based on reciprocity of operations. This concept in many countries preclude a more rational assessment of air services, one which transcends narrow sectoral interests of an air carrier and provides the optimum overall benefits from tourism, trade and investment for the economy, protection of consumer interests and enhancement of competition.

Another anomaly in aviation governance, particularly from a performance point of view, is that we are right now not planning for oncoming growth the way we should, particularly if we increase connectivity according to our global mandate. Even as we are, and taking into consideration current growth patterns *Airbus Industrie*, in its forecast from 2011 to 2030 envisions the need for more than 26,900 passenger airliners with seating capacities of 100 seats and above, along with over 900 new factory-built freighter aircraft. In the same timeframe, the world's overall passenger aircraft inventory will more than double from today's 15,000 to more than 31,500 by 2030. Boeing in its forecast for 2011 to 2031 states that the long-range forecast for 2011 anticipates delivery of 33,500 new airplanes over the next 20 years, valued at more than $4.0 trillion. Retrospectively, Boeing's long term forecasts have tended to be conservative compared to actual industry performance, but well thought through.

The figures of both these "predictions" would cater mostly to the megacities of the world (26 in number, according to the IATA Report), which have a population of more than ten million and which account for more than 20 % of global air travel. Sixty-two cities, with more than five million population will account for another 40 %.

Although at first glance this rosy picture seemingly suggests that air transport is progressing well, there is something wrong in this scenario. And that is that we would be perpetuating the "predict and provide" principle that we have followed over the past 67 years of the formal existence of commercial air transport if we do not decide what to do about this growing and increasing trend. What is wrong is that, while the "predict" part of the equation is fine, the "provide" part of it, if blindly catered to without looking at the systemic consequences, would lead us to all sorts of problems, particularly in the environmental field and human resources field.

The 1983 principle of predict and provide for transport planning, advocated by the Economist Adams, was a disaster for the simple reason that it took as inevitable perpetual transport growth at face value. We should "predict and plan".

Let me sum up as to why we cannot, under the present circumstances, predict and plan:

- All scheduled international commercial air transport services are prohibited except to the extent they are permitted;
- All bilateral and open skies agreements are reciprocal and subject to the nuances of aeropolitics and protectionism and arbitrary demarcations of market share;
- Therefore airlines do not have freedom to access of markets;
- There are rigid and archaic ownership and control regulations governing so called "national carriers";
- In many instances, this effectively precludes direct foreign investment in airlines;
- All the above unduly prevent connectivity, which is the meaning and purpose of meeting the needs of the people of the world for regular, efficient and economical air transport.

As the foregoing discussion showed, air transport is a different animal from other aspects of trade, in that it is circumscribed by the need for airlines to obtain permission of the grantor State to obtain market access. In this context, how does one approach market access in the coming decades? At the most fundamental level, the advantages of free trade as would apply to air transport would be that they would encourage States to trade freely with their trading partners which would in turn help in the growth of the global economy; they would give the consumer a better choice of products and competition generated by free trade would bring down the price of the product. Arguments against free trade in air transport would be that globalization and liberalization will take jobs away from a State; the limit of imports would keep money in the State; free trade could be a threat to national security and a State could develop dependence on the expertise of other more advanced States. Free trade increases national wealth and promotes foreign investment, both of which are absent in the present structure of market access in many States.

The main consideration, leading up to efforts by the international aviation community to achieve a deregulated global airline industry, is involved with the question as to whether free market principles can be applied globally to air transport. What needs to be considered is whether we are ready to accept the throwbacks as consequences of free market competition in air transport, particularly in losing national prestige projected by flag carriers. One of the corollaries to industry deregulation is the introduction of free market competition when companies switch from operative performance to competitive performance. Competition therefore emphasizes the need to focus on a company's performance in relation to its competitors. This principle can be readily apply to various industries that have already been deregulated, such as the motor vehicle industry, chemical industry and information technology industry. The operative question is "are these good analogies for application to the air transport industry?" Whatever be the answer to this question, if the deregulated domestic air transport industry of the United States were to be considered an analogy, one could say that a deregulated system in the United States, introduced in 1978, has led to a more efficient airline system in the country. Whatever be the case, access to facilities in a competitive market is essential toward attaining fluidity of market forces. In the air transport industry, this can be translated to mean that if free markets do not exist in the supply of complementary facilities, there will be no positive impact of liberalization. The complementary services in the supply of air transport are airport access, computer reservation systems and airport and air regulation services.

The International Chamber of Commerce (ICC), in a policy statement has expressed the view that the efficiency of air transport would be enhanced by creating more open markets and more flexibility with regard to foreign ownership. Given air transport's capability to facilitate economic activity, its liberalization would enable the sectors that make use of it to become non efficient. ICC was in favor of a freer exchange of air services throughout the world and is convinced that it is time to move beyond the existing bilateral system, toward a genuine multilateral liberalization of air transport of course, liberalization would give way to competition, which in turn would impel airlines to pool their resources in order that they maximize on such assets as code sharing and airport slots. However, alliances do not necessarily mean lack of competition between partners. Airlines within alliances have to do their utmost to gain market access and keep their businesses alive. In order to do this both private enterprises and the States in which these enterprises are entrenched have to be equally competitive.

Any agreement to bring in an aspect of trade within a liberalized framework is generally a pro-active measure, which brings to bear the willingness and ability of the governments to face trading issues squarely in the eye. However, any agreement for trading benefits would be ineffective without the element of competition, both between enterprises and between States. The essential requisite for success in trading relations is competition, which in turn leads to national prosperity. A free trade agreement is merely the catalyst in the process.

So what do we do in aviation to get back on track with our strategic direction and achieving our main objective—providing the consumer with the best possible connectivity in an efficient and economical manner? Forecasting the fate of the air transport industry is dicey at the best of times. No one knows the future. The need for air transportation between major cities continues to grow, as populations grow and prosper and economic wealth develops rapidly. On this basis, I believe one can plan. Here are my suggestions:

(a) We should conduct research into all aspects of air transport which are of international importance. This is provided for in the Chicago Convention. Such studies, taking into account global, regional and national economic trends, could analyse their effects on the demand for air transport and how such demands could be met. This could result in a compendium of planning for States, aircraft and component manufacturers, environmentalists and service providers.

(b) One of the forgotten issues in air transport—the right of the consumer (particularly the passenger) should be a key issue for study. The October 2011 crew strike of QANTAS as well as the August/September 2012 strike by Lufthansa crew left thousands of passengers stranded at airports. Although these grave consequences to passengers were the result of industrial actions, States must take responsibility for providing regular, unbroken air transport that is not arbitrarily hindered by unmet demands of the service provider. This is also emphasized in the Chicago Convention. There should be core guidance to States with a view to ensuring regularity of service. This would meet the objective that air transport meets the needs of the people for air travel.

(c) The Chicago Convention also mentions studying issues affecting international air transport, including ownership and control issues as well as air transport services on trunk routes and the development of plans. We should produce a comprehensive annual or triennial report on global trends and their effect on air transport, with guidance to key stakeholders in the provision of air services that meet the needs of the consumer while ensuring sustainable development of air transport. This will de-fragment the current process whereby each key player has his own forecast—often based on a "predict and provide" model rather than a strategic management model.

(d) We should discuss justifications for and against free trade in aviation when considered against national concerns on aviation safety and security; aircraft manufacturing and commercial aspects concerned therewith including government subsidies and the manner in which civil aviation would affect trade surpluses and deficits. Another aspect worthy of in depth study is the effect on the global and national economies of open skies.

Creating and preserving understanding among the people of the world is not just about visiting the Eiffel Tower or the Coliseum, although no one doubts the fact that air transport is structurally interrelated to tourism. But at its core, air transport is about serving the travelling public in the most efficient and economical way, while preparing for what is to come. I can only think of what Edward Warner, the United States Representative said at the Chicago Conference. He said:

Our first purpose will be to smooth the paths for civil flying wherever we are able. We shall seek to make it physically easier, safer, more reliable, more pleasant; but I believe it will be agreed also that we should maintain the constant goal that civil aviation should contribute to international harmony. The civil use of aircraft must so develop as to bring the peoples closer together, letting nation speak more understandingly unto nation.

Dr. Warner had notably stressed on the purpose of civil aviation to be the promotion of international harmony and dialogue between nations. He had also made it clear that the seminal task of civil aviation is to bring the people of the world together through understanding and interaction. It is clear that at this stage at least, civil aviation was recognised more as a social necessity rather than a mere national or economic factor aligned towards protectionism.

Once we get our strategic direction right, perhaps we could discuss performance governance in air transport in a more meaningful manner.

References

Abeyratne R (2012) Aeronomics and law: fixing anomalies. Springer, Heidelberg
Shaw MN (2003) International law, 5th edn. Cambridge University Press, Cambridge

Chapter 6
Airlines: The Other Side of the Coin

6.1 Airlines and Connectivity

"Connectivity"—the most important buzzword in meeting the needs of the air transport consumer—does not only bring to bear the significance and involvement of rules and regulations issue but also highlights the key player in delivering the air transport product—the airline and its economics of operation. The Air Transport Annual Report 2012 of the World Bank states that air transport continued its global recovery in the fiscal year 2012, and expected a collective profit for the year. However, according to the World Bank forecast airline these profits would be less than half of the $8.4 billion the industry earned in 2011. High oil price and low global recovery were the causes for this decrease. The World Bank forecast for the next two decades is consistent with that of ICAO, which has stated that the outlook is positive with a strong growth rate and a doubling of the annual passenger carriage by 2030.

The airlines are key to meeting the needs of the travelling public through optimum connectivity, the availability of services and value for money. They are represented by IATA which acts primarily as their lobby that pursues their commercial interests. At ATConf/5 IATA made two interventions on the fundamentally important subjects of ownership and control and slot allocation, both of which are crucial for serving the travelling public. Prior to discussing IATA's suggestions, it is to be noted that, on ownership and control, Members of the Latin American Civil aviation Commission (LACAC) observed that although liberalization of air transport has been debated over the decades, no acceptable solution that would appeal to the majority of the airlines had been reached. LACAC recommended that the "flags of convenience" issue should be given more attention, particularly in the emergence of carriers established under the flags of convenience practice. Other issues that demanded attention were the deterioration of safety and security standards, possible flight of capital; impacts on labour, national emergency requirements and assurance

R. Abeyratne, *Regulation of Air Transport*, DOI 10.1007/978-3-319-01041-0_6,
© Springer International Publishing Switzerland 2014

of service and, in the long run, anti-competitive effects from industry concentration.[1] IATA suggested four steps to liberalize airline ownership and control: distinguish between commercial control conferred by ownership and regulatory control exercised by the licensing authority; remove restrictions on ownership; make regulatory control the responsibility of the designating State(s); and provide control of safety and security through the adoption and implementation of the relevant ICAO/ECAC Model Clauses.[2]

On the issue of slot[3] allocation, IATA emphasized that existing IATA schedule coordination provided flexible and fair guidelines for slot allocation on a global basis and that national or regional rules could unbalance or upset the system, particularly if they happen to be inconsistent with the global system. Therefore IATA suggested that any slot allocation system should respect global compatibility towards the maximum use of airport capacity through a market driven approach. IATA further suggested that slot allocation should be transparent, fair and non-discriminatory, as well as simple, practical and economically sustainable.[4]

A slot is an intangible asset representing an interest. It is similar to the property rights represented by: a share certificate or a professional licence, and derives value not from its intrinsic physical nature but from what it represents. As such, it is essential to define the various rights that make up a slot. The existing slot allocation system and, to some extent, trading has confined slot rights solely to runway usage. A clear study demarcating the benefits of formally incorporating rights beyond runway capacity may not yield a clear understanding of the subject would require an involved and carefully thought through analysis.

---

[1] ATConf/5-WP/99.

[2] ATConf/5-WP/26.

[3] Slots have been defined in EC Regulation 95/93 as: "The scheduled time of arrival or departure available or allocated to an aircraft movement on a specific date at an airport co-ordinated under the terms of this Regulation". Other definitions exist. For example, "The scheduled time of arrival or departure available for allocation by, or as allocated by, a co-ordinator for an aircraft movement on a specific date at a fully co-ordinated airport"—IATA Worldwide Scheduling Guidelines. These guidelines, widely used by slot co-ordinators (and which create preference rights largely consistent with the EC Regulations) do not appear to have legal status in Regulation 95/93 other than an obligation on the co-ordinator to "take into account" such guidelines "The specific time allocated for an aircraft to land and to take-off"—ICAO's International Civil Aviation Vocabulary.

[4] ATConf/5-WP/27. It should be noted that Article 15 of the Chicago Convention provides *inter alia*: "Every airport in a contracting State which is open to public use by its national aircraft shall likewise, subject to the provisions of Article 68, be open under uniform conditions to the aircraft of all the other contracting States. The like uniform conditions shall apply to the use, by aircraft of every contracting State, of all air navigation facilities, including radio and meteorological services, which may be provided for public use for the safety and expedition of air navigation. Article 68 provides: "Each contracting State may, subject to the provisions of this Convention, designate the route to be followed within its territory by any international air service and the airports which any such service may use.

A study released in 2001 stated:

> There is considerable concern that existing allocation procedures for airport slots produce inefficient outcomes and competitive distortions. Transition to a market-based allocation system with appropriate safeguards against concentration of slots would improve efficiency, encourage competition and yield significant benefits for consumers. Adaptations of auction formats used for radio spectrum allocation provide a feasible means of allocating airport slots. This would need to be complemented by a formal secondary market.[5]

The Study also mentioned that current slot allocation procedures perform inefficiently as they did not allocate slots with the objective of generating greatest benefit for consumers and the economy at large. The current system of allocating slots on the basis of historic precedence was, the Study mentioned, inconsistent with obtaining the greatest possible benefit from available airport capacity. It is possible that a potential user could generate much greater social benefit from a slot than its existing user, but the slot would nevertheless be allocated to the existing user.[6]

A contemporaneous study reflected:

> For a variety of reasons, airport access is not priced at market clearing levels and thus "slots" are scarce and have an economic value. However, unlike many other valuable resources, no clear tradable rights have ever been universally defined or formally allocated. As the value of UK slots (and hence the cost of any misallocation) is likely to increase in the coming years, the introduction of a legitimised secondary market offers a logical way to let the market determine how slots – the vast majority of which are grandfathered at present at congested airports such as Heathrow and Gatwick – should be held[7]

The same study said that the challenges of clarifying "ownership" for the purposes of the primary allocation of slots need not prevent the definition of tradable rights in slots and legitimising a secondary market. In a competitive market slots ought to move to those who value them the most irrespective of initial allocation. This suggests the issue of "ownership" is a separate policy decision relating to the distribution of scarcity rents that ought to be distinguished from the objective of maximising the value of production from slots. In the transfer of slots from one airline to another, be it on exchange, sale or auction, evidence that clear tradable rights are necessary for an efficient secondary trading system is a must, notwithstanding the fact that some secondary trading already occurs despite no clearly defined legal rights. There has essentially to be a clear definition of legal rights so as to obviate the ambiguity, uncertainty and risk presently surrounding grey market trading. If such rights are clearly defined a firmer basis will be provided upon which private bargains may be struck and efficiency improvements in airport capacity allocation realised.

---

[5] *Auctioning Airport Slots – A Report for HM Treasury and the Department of the Environment*, January 2001, at IV.

[6] *Id.* V.

[7] The Implementation of Secondary Slot Trading, November 2001, Civil Aviation Authority, CAA House, 45–59 Kingsway London WC2B 6TE, at VII.

In late 2012 India began considering introducing an auction for peak-time landing slots for airlines operating into in a shake-up of aviation policy under the government's economic reform drive.[8] In 2011, taking into consideration the looming airport congestion problem in Europe, the European Union (EU) proposed a market based system for slot allocation. The proposed measures on slots would allow airports across the union's 27 member states to handle 24 million more passengers a year by 2025, generating 5 billion Euros (about $6.67 billion) in additional economic activity and creating up to 62,000 jobs.[9] At a meeting in London on 19 March 2012, the House of Lords EU Sub-Committee on the Internal Market took evidence from British Airways, easyJet the European Regions Airline Association and other organisations as part of its examination of these proposals. Prior to 2008, slots in Europe could be exchanged, but not sold. In April 2008, the EU declared that existing EU legislation on airport slot allocation did not prohibit so-called "secondary slot trading," bringing an end to years of discussion. In its statement the European Commission stated that "the Commission believes that prohibiting such secondary trading causes the number of slots available to new entrants – and to incumbent air carriers seeking to extend their services – to be lower than it could be. An air carrier may retain a slot even when its market value far exceeds the value that the air carrier generates from retaining and using the slots".[10]

ATConf/6 discussed connectivity in a holistic manner, recognizing that there is no single definition of air transport "connectivity". For purposes of reaching some understanding of connectivity, it is considered a property of a network and can be defined in such a way as to constitute an indicator of the network's concentration. Therefore, connectivity is the ability of a network to move a passenger from one point to another with the lowest possible number of connections and without an increase in fare, focusing on, from a commercial perspective, minimum connecting times with maximum facilitation ultimately resulting in benefits to air transport users.

In past years, the pace of liberalization has been affected by the global economic recession, which has led to a reversion toward trade protectionism in many countries. Air transport has not escaped this setback, even though in practice liberalization could be a stimulus to recovery. As of today, the vast majority of arrangements are under this type of bilateral Air Services Agreement, and the primary focus is on removing national air carrier ownership and control provisions

---

[8] http://ibnlive.in.com/news/government-mulls-airport-slot-auctions-in-new-air-policy/296249-3.html.

[9] http://www.nytimes.com/2011/12/01/business/global/europe-to-propose-a-new-system-to-allocate-airport-slots.html?_r=0. The Commission backed down from proposing that new airport capacity be allocated by auction. Currently, when an airport expands, additional take-off and landing slots are allocated by an independent coordinator, which sets aside 50 % of them for new entrants. The other 50 % goes to incumbent airlines on a first-come-first-served basis.

[10] http://www.euractiv.com/transport/eu-allow-sale-airport-slots-news-219776.

from air services agreements and agreeing on acceptance of ownership and control provisions based on principal place of business in one or more of the participating States. Therefore, the key challenge is linked to bilateral restrictions which limit the availability of services for the air transport user. According to the industry, airline liberalization can further increase demand and ensure that the services, which are providing increased connectivity, are sustainable over the long-term. It provides the commercial freedom necessary for airlines to adjust capacity appropriately to meet changes in air travel demand. By way of example, the growth in air services between Poland and the United Kingdom (UK) since 2003 resulted in a gross domestic product (GDP) increase of 27 % for Poland, whereas the increase in the already well-served UK was a much smaller 0.5 %. These changes provide an estimated long-term boost to Poland's GDP of US$634 million per annum. The UK also benefited, with an estimated boost to its GDP of US$45 million per annum.

Network carriers generally offer scheduled flights to major domestic and international cities while also serving smaller cities; the carriers normally concentrate most of the operations in a limited number of hub cities, serving most other destinations in the network by providing one-stop or connecting service through the hubs. An efficient utilization of the hub allows airlines to offer better connectivity. The "hub and spoke" model is a system which enhances efficiency in transportation by greatly simplifying a network of routes. Many airlines supplement the "hub and spoke" model with codeshares, partner flights, or a small commuter airline. It should be noted that the way in which airlines price tickets can also impact connectivity, notably in the case of transit by flight stage; if a trip is sold by flight stage as opposed to origin to destination (i.e. two tickets rather than one ticket), there can be significant increases in transit times, hence a loss of connectivity.

At ATConf/6 IATA, with other organizations, advised that competition in the airline industry remains strong. In Singapore for example, 90 % of passengers travel on routes operated by two or more carriers. In Europe, the number of routes with two or more carriers tripled between 1992 and 2009. In many domestic and regional markets, the introduction of new model carriers has intensified competition. New model carriers currently account for 50 % of seat capacity in India, and 20 % of seat capacity across Asia-Pacific.

6.2   Charges and Taxes: A Burden on Airlines

ATConf/6 addressed the issue of charges and taxes and their impact on airlines and the consumer of air transport. IATA and Airports Council International (ACI) presented a joint paper[11] which said, inter alia:

---

[11] ATConf/6-WP/75.

There is increasing concern with the growing proliferation of taxes beyond the scope of ICAO policies levied on international air transport. Although the industry fully respects the right of autonomous States to impose taxes, excessive taxation on international air transport for the sole purpose of generating State revenues has a severe negative impact on the global economy, the national economic welfare and endangers the sector's economic recovery. Moreover, such taxes are socially not sustainable as they usually affect lower income travellers in a disproportionate manner.[12]

A tax is a "pecuniary contribution made by persons liable, for the support of government".[13] Courts have adjudicated that a tax is "a pecuniary burden laid upon individuals or property to support the government and is a payment exacted by legislative authority". It has also been identified as "annual compensation paid to government for annual protection and for current support of government." An early American decision identified a tax as:

> A ratable portion of the produce of the property and labour of the individual citizens, taken by the nation, in the exercise of its sovereign rights, for the support of government, for the administration of the laws, and as the means for continuing in operation the various legitimate functions of the State.[14]

According to these definitions, a tax is a very general imposition, often described as a "once and for all" payment. Therefore, a tax could not be named, as a specific tax, such as "aviation fuel tax" or "aircraft equipment tax". The fact that a tax was levied "for the support of the government" makes its general nature more explicit.

A tax is a "pecuniary contribution made by persons liable, for the support of government".[15] Courts have adjudicated that a tax is "a pecuniary burden laid upon individuals or property to support the government and is a payment exacted by legislative authority". It has also been identified as "annual compensation paid to government for annual protection and for current support of government." An early American decision identified a tax as:

> A ratable portion of the produce of the property and labour of the individual citizens, taken by the nation, in the exercise of its sovereign rights, for the support of government, for the administration of the laws, and as the means for continuing in operation the various legitimate functions of the State.[16]

According to these definitions, a tax is a very general imposition, often described as a "once and for all" payment. Therefore, a tax could not be named, as a specific tax, such as "aviation fuel tax" or "aircraft equipment tax". The fact that a tax was levied "for the support of the government" makes its general nature more explicit.

In the 1956 case of *Heirs* v. *Mitchel*,[17] the court held that a tax was:

---

[12] *Id.* at 2.

[13] Black's Law Dictionary, 1951.

[14] *New London* v. *Miller*, 1941 *Connecticut Reporter* at 112.

[15] Black's Law Dictionary, 1951.

[16] *New London* v. *Miller*, 1941 *Connecticut Reporter* at 112.

[17] 1956 *Southern Reporter* at 81.

An enforced contribution of money or other property, assessed in accordance with some reasonable rule or apportionment by authority of some sovereign State on persons or property within its jurisdiction for the purpose of defraying the public expenses. Therefore, a tax came to be known as a "contribution" and was regarded in a general sense to be any contribution imposed by government upon individuals, for the use and service of the State, whether under the name of toll, tribute, tallage, gable, impost, duty, custom, excise, subsidy, supply, aid or *any other name.*

The legal definition of a tax is that it is an enforced contribution by the public or section thereof, introduced by legislative decree, for the purposes of defraying public expenses. Judicially, a tax has been identified as a 'contribution', among other synonyms, including, quite disturbingly, with 'any other word'.[18] Experts in taxation maintain that the 'efficiency' test in taxation calls for devising tax levies which cause minimal reduction in or disruption of, overall productivity of a society. It is in this perspective that the overall context of taxation in the field of international air transport should be viewed.

In many instances "taxes" imposed on international air transport have been labeled as iniquitous.[19] It is strongly claimed that a tax which is "a compulsory contribution levied upon persons, property or business for the support of government: any assessment"[20] is an onerous demand upon any one's person or resources and when imposed upon international air transport justifies its definition as a verb— to subject to a severe strain. While it is accepted that taxation must be for a public purpose the amount of the tax charged must be compatible with principles of commerce and should be proportionate to the cost of the specific facility or services used rather than the cost of overall governmental services in general. The formula must admit of the tax being directly proportionate to the cost of the service or facility used.

6.2.1 Definition of Charge

A "charge" has reference to impositions for improvements which are specially beneficial to particular individuals or property, and which are imposed in proportion to the benefits supposed to be conferred. Charges are special and local impositions upon property in the immediate vicinity of municipal improvements and are laid with reference to the special benefit which the property is expected to have derived therefrom.

---

[18] *In re. Mytinger*, D:C: Tex., 31 *F. Supp.* 977 at 979.
[19] See *Aviation Daily*, 16 November 1990 at 328 where IATA Director General calls the $1 "Facilitation Fee" proposed by the US and which was to be imposed on aliens visiting the US a "tax". See also, *Aviation Daily*, 11 January 1990 at 71 where he makes a strong plea to US Secretary of Transport to urgently review the taxation of international transport. This tax was later withdrawn by the United States authorities, largely as a response to the effective lobbying against the imposition of this tax by such organizations as IATA. See Abeyratne (1993), at pp. 450–460.
[20] 7See generally, Hinshaw (1939) at pp. 75–94.

A charge levied upon the products of a particular industry is expected to be utilized in the improvement of that industry, while a tax is generally imposed in the national interest and is directed accordingly towards the national treasury. In concept, the former is not objectionable, since it is calculated to benefit a particular industry for which the charge is collected, while the latter, it is claimed, should be borne by States as part of their national responsibility. However, this clear demarcation has often been shrouded in anomalous terminology resulting in a passenger service charge being identified as a tax imposed for the national benefit.

In a broader sense, taxes, as have been judicially defined, could be considered as including assessments and charges. But practically, a "tax" is a public burden imposed generally on the inhabitants of a State or upon a division thereof for governmental purposes, without reference to particular or peculiar benefits to particular individuals or property. The main criticism of a tax is that it is a compulsory contribution levied upon persons for the support of government and therefore is a heavy demand upon one's person or resources.

Article 15 of the Chicago Convention enshrines the basic philosophy of charges on airports and air navigation services. The provision states *inter alia* that every airport in a Contracting State which is open to public use by its national aircraft shall likewise be open under uniform conditions to the aircraft of all the other Contracting States. The like uniform conditions shall apply to the use, by aircraft of every Contracting State, of all air navigation facilities, including radio and meteorological services,[21] which may be provided for public use for the safety and expedition of air navigation. Article 15 also provides that any changes that may be imposed or permitted to be imposed by a Contracting State for the use of such airports and air navigation facilities by the aircraft of any other Contracting State shall not be higher:

(a) As to aircraft not engaged in scheduled international air services, than those that would be paid by its national aircraft of the same class engaged in similar operations; and
(b) As to aircraft engaged in scheduled international air services, than those that would be paid by its national aircraft engaged in similar international air services.

Article 15 subsumes three fundamental postulates:

(a) Uniform conditions should apply in the use of facilities provided by airports and air navigation services;
(b) Aircraft operators should be charged on a non-discriminatory basis; and
(c) No charges should be levied for the mere transit over, entry into or exit from the departure of a Contracting State.

[21] Article 28 of the Chicago Convention calls on each Contracting State, so far as it may find practicable, to provide airport and air navigation facilities, in accordance with the standards and practices recommended or established in pursuance of the Convention.

The ICAO Contracting State, which has notified its adherence to the Chicago Convention, could exempt itself from Article 15, which binds Contracting States to adherence under customary international law. The binding force of the Chicago Convention on its Contracting States is ensured by the Vienna Convention on the Law of Treaties,[22] which recognizes treaties as a source of law, accepts free consent, good faith and the *pacta sunt servanda* as universally recognized elements of a treaty.[23]

An elaboration of the basic principles of Article 15 in the context of airports and air navigation services charges is contained in *ICAO's Policies on Charges for Airports and Air Navigation Services*[24] which call for an equitable and fair determination and sharing of airport and air navigation services costs—a principle which has had sustained application by Contracting States. The Council Statements apply to charges as referred to in the provision of airports and air navigation services within the parameters of Articles 15 and 28 of the Chicago Convention. In this context, charges are levies calculated to defray the costs of providing facilities and services for civil aviation. Charges are diametrically opposed in concept to taxes which are levies directed at raising national and local government treasury funds to be used for non-aviation purposes.[25]

In recent years, a compelling need has arisen to address in greater detail the basic principles pertaining to cost recovery applicable to both airports and air navigation services. This is because, primarily, the organizational structure of airports and air

---

[22] *Vienna Convention on the Law of Treaties*, U.N. Doc A/CONF.39/27 (23 May 1969).

[23] The *pacta sunt servanda* which is enshrined in Article 26 of the Vienna Convention provides that every treaty in force is binding upon the parties and must be performed by them in good faith. The validity of a treaty or of the consent of a State to be bound by a treaty may be impeached only through the application of the Vienna Convention, which by Article 42.1 requires that one of the following circumstances exist: that a treaty be derogated only in circumstances the treaty in question so specifies; that a later treaty abrogate the treaty in question; and that a *novus actus interveniens* or supervening act occur making the performance of the treaty impossible. See Abeyratne (1996) at pp. 11–20.

[24] *ICAO's Policies on Charges for Airports and Air Navigation Services*, Doc 9082/6, Ninth Edition:2012 at 1. There have been a few instances of litigation in the United States on charges imposed by airlines. In the 1961 case of *Aerovias Interamerivacanas de Pananma, S.A.* v. *Board of County Commissioners of Dade County Florida* (197 F.Supp. 230) the US Supreme Court ruled that the plaintiffs, who sued the defendants for charging them airport fees in excess of those charged with regard to the national carrier PANAM, were entitled to cite and obtain benefit of the principle in Article 15 of the Chicago Convention—that airport charges should be equally applied to all carriers alike who operate air services into and out of an airport. In the 1994 case of *Northwest Airlines* v. *County of Kent* (510 U.S. 355), seven commercial airlines, petitioners in the case, claimed that airport user fee charges imposed on them were unreasonable and discriminatory. The airport in question was owned by Kent County. The airport charged its users: airlines both commercial and non-commercial and non aeronautical businesses such as concessionaires, car rentals based on a "cost of service" accounting system known as the Buckley methodology. . The court found the airport liable on the ground that it was erratic in its charging policy since it undercharged general aviation and overcharged the concessionaires.

[25] See Abeyratne (1991) at pp. 106–118. See also Abeyratne (1995) at pp. 48–59.

navigation services providers have changed in many parts of the globe into autonomous and privatized bodies. Earlier, when the majority of these service providers were instrumentalities of State or branches of government, the charges policies of ICAO, as contained in the Chicago Convention and the Council Statements could be applied within their basic philosophy.

Now, however, with States having divested themselves of this responsibility and vested the operator of their airports and air navigation services in privatized and other autonomous entities, cost recovery policy may need to be reviewed. With this in view, ICAO suggested to the Conference on the Economics of Airports and Air Navigation Services (ANSConf 2000) held in Montreal on 19–28 June 2000, that the profile of basic cost recovery policy may need to be raised.[26] This measure would still be adopted within the parameters of existing policy calling for revenues from charges levied on international civil aviation and shall only be applied towards defraying the costs of facilities and services provided for international civil aviation. It was also recommended that revenues from other sources than charges on air traffic shall be taken into account before the cost basis for charges on air traffic is determined. ICAO also advised ANSConf 2000 that airports and air navigation services may produce sufficient revenues to exceed all operating costs and so provide for a reasonable return on assets to contribute towards necessary capital improvements. Of course, the governing principle would be that consultation with users shall take place before significant changes in charging systems or levels of charges are introduced.[27]

ICAO's recommendations to ANSConf 2000 were both timely and practical, given the evolving fabric of economic forces which now govern airports and air navigation services. The recommendations also stimulate some reflection on the complexities of financing principles now applicable to the services provided by airports and air navigation services providers. In substance, the issue of costing and pricing of services would be dependent upon underlying practices and economic principles.

The theory of marginal cost pricing admits of fixing a price equivalent to the cost of producing one additional unit.[28] Translated to the context of airports and air navigation services, it would mean that marginal cost is that cost incurred in providing an additional unit of output. Although an efficient allocation of resources can be optimally achieved by a pricing policy when the price of goods or services is set equal to the marginal cost of providing such goods or services, the economic rationale of marginal pricing is essentially based on the principle that only users who would value a service as much as its cost of provision will be prepared to pay for using the service.

In order to study the applicability of marginal cost pricing to the provision of airports and air navigation services, ICAO appointed in 1999 a Secretariat Study

---

[26] *ANSConf-WP/4* at para. 5.1.

[27] *Id.* para. 5.3.

[28] Boiteux (1964) at p. 59.

Group composed of members from factors as the bunching of aviation and non-aviation revenues and their effect on the overall pricing policy relating to airports and air navigation services and a significant paradigm shift from Article 15 of the Chicago Convention and the Council Statements.

6.2.2   Marginal Cost Pricing

The suggestion, that marginal cost pricing be considered in setting charges, was rejected at an earlier ICAO Conference held in 1991.[29] Subsequently, the Air Navigation Services Economics Panel (ANSEP) of ICAO addressed the possibility but rejected it on the grounds that such a practice would not be consistent with the equity principles enunciated by Article 15 of the Chicago Convention. Current policy on cost pricing of airports and air navigation services can be attenuated from the ICAO *Manual on Air Navigation Services Economics*[30] which, under paragraph 5.11 provides that unit cost is determined by dividing the cost base by the estimated charging units, making the charges structure accomplish the averaging out of costs among the various users, irrespective of the costs that individual users impose. In practice, the implementation of ICAO's pricing strategy of the recovery of cost through a process of non-discrimination has led to cost pricing combined with factors determining the user's ability to pay based essentially on the weight of the aircraft for which services are rendered. The Study Group concluded that the application of economic principles within the parameters of the equity and non-discrimination principles of Article 15 was permissible, noting the inherent difficulties that would be posed by the application of principles of marginal cost pricing, owing to the complex calculations involved. The Group suggested that such an application should remain optimal, perhaps to be used only in instances where airports and air navigation services were provided under congestion and space constraints. The Group's conclusion was that marginal cost pricing, owing to its complex nature, should be applied with caution.

The recommendation of the Study Group of the ICAO Secretariat, that principles of marginal cost pricing be applied with caution, should be given serious consideration in the context of civil aviation, particularly in applying the attributes of marginal cost pricing to aviation. Simply put, if a service provider incurs a certain cost in providing services for regular—flow traffic, any additional traffic that may flow requiring the services of that provider would incur additional marginal costs which such additional traffic may have to bear. In other words, if total costs for the provision of regular services at a fixed cost basis were to be represented by 'p' and the regular flow of aircraft were to be represented by 'q', each user would pay as a cost based price varying of course with the weight of each aircraft. If an additional

---

[29] Conference on Airport and Route Facility Management (CARFM) 1991.
[30] Doc 9161/3.

flow, representing '*r*' were to cost an additional price of '*s*', the user causing '*s*' would have to pay. Since the average marginal costs borne by the additional flow would be greater than the cost borne by the regular flow. This would particularly prove iniquitous if an aircraft of lesser weight were to pay more than an aircraft of higher weight. Although, as the Secretariat pointed out to ANSConf 2000[31] that such pricing policy may achieve the objective of encouraging air traffic to move from congested peaking hours to periods of less peaking or no congestion, the International Air Transport Association (IATA) pointed out that marginal cost pricing is concerned with current and future costs and not with past—sunk costs which do not apply to an additional user.[32] IATA also averred to the distinction between short-run and long-run marginal cost pricing cautioning that long-run marginal cost pricing will pose difficulties with regard to amorphous long-term infrastructural costs (such as incurred for the construction of terminals and runways over a period of time).

Being mindful of these difficulties, but at the same time considering the need for and value of additional work on this area to cover all grounds, the Secretariat has suggested to ANSConf that ICAO policy be reworded in the Council Statements to read as follows:

> The costs should be determined on the basis of sound accounting principles and may reflect as required *other economic principles* (my emphasis) provided these are in conformity with Article 15 of the *Convention on International Civil Aviation* and other principles in the present document.

The ICAO Secretariat also advised ANSConf 2000 of text inserted in the *Manual on Air Navigation Services Economics* (Doc 9161/3) in paragraph 5.7 to the effect that the application of economic principles to setting charges should emphasize the need to recover costs in an equitable manner from the use of air navigation services. The two governing criteria are the cost of providing air navigation services and the effectiveness of services rendered. Paragraph 5.8 stipulates that the aim of an economic approach to setting charges is to bring the demand for and the offer of air navigation capacity into equilibrium so that the provision of additional capacity is encouraged. The level of charges should be considered in performing the required investment appraisal (including cost/benefit analysis). An example of an appropriate economic approach would be to apply a two part tariff with a fixed rate for "entry" into the system and a variable element to reflect the extent of the service rendered. It was also contended that, in order to apply economic principles to setting charges, it would be necessary to identify costs associated with the air navigation services components being charged for.

The Secretariat subsumed ICAO philosophy by quoting text inserted in paragraph 5.6 of the *Airport Economics Manual* (Doc 9562) that the application of economic principles to setting charges which are consistent with the Council

---

[31] *ANSConf-WP/14* at p. 2.

[32] *ANSConf-WP/80* at p. 2.

Statements in Doc 9082/5 should emphasize the need to recover costs in an equitable manner from the users of airport services.[33]

ANSConf 2000 consensually agreed that States should not be restricted to resorting to a single principle in setting their airport and air navigation charges. An approach best suited to States, in terms of their own local circumstances was acceptable, according to the Conference, provided such an approach met internationally-accepted standards and principles. Consideration was given to the significance of demand factors and the deleterious effect of abuse of market power if advanced economic principles on pricing were to be set arbitrarily and contrary to the principles of equity and non-discrimination as enunciated in Article 15 of the Chicago Convention. Accordingly, the Conference made two recommendations, the first being that a suitable amendment be made to the Council Statements in Doc 9082 to the effect that charges should be determined on the basis of sound accounting principles and may reflect, as required, other economic principles, provided that these are in conformity with Article 15 of the Chicago Convention and other relevant principles in Doc 9082.[34]

The second recommendation of the Conference called upon States to urge their service providers that adopt economic pricing principles to report details on both the level and allocation of fixed and variable costs, and to consult with users on how marginal costs are estimated and used in setting charges. In this regard, ICAO is requested to introduce supplementary guidance aimed at ensuring that the use of economic pricing for capacity management does not result in over recovery of costs while at the same time providing for a reasonable return on investment. ICAO is also charged with the task of examining whether the traditional approach to measuring discrimination and cross-subsidization needs to be revised given the increased interest in applying economic principles in setting charges.[35] These recommendations were made based on suggestions of the ICAO Secretariat[36] and Airports Council International.[37]

6.2.3 The Single Till

The "single till" concept, which admits of both aeronautical and non-aeronautical revenues being used for aeronautical purposes, brings to bear the need to determine whether airports, as autonomous business entities, should as of right use non

---

[33] See Appendix, *ANSConf-WP/14*.

[34] Report of the Conference on Economics of Airports and Air Navigation Services—Air Transport Infrastructure for the twenty-first century, Montreal, 19–28 June 2000, Doc 9764, ANSConf 2000, Recommendation 19.

[35] *Id*. Recommendation 20.

[36] *ANSConf-WP/14*.

[37] *ANSConf-WP/54*.

aeronautical revenues for purposes other than developing airport and air navigation services for the aircraft they serve, or whether both aeronautical and non-aeronautical revenues should be put back into one single till for reducing charges on airport that serve those airports. The cost basis for airport charges setting is set out in ICAO policy as follows:

> The cost to be shared is the full cost of providing the airport and its essential ancillary services, including appropriate amounts for cost of capital and depreciation of assets, as well as the cost of maintenance and operation and management and administration expenses, but allowing for all aeronautical revenues plus contributions from non aeronautical revenues accruing from the operation of the airport to its operators.[38]

This provision envisages the pooling of airport revenues in one single source or "till" whether they be aeronautical or non-aeronautical revenues.

The underlying purpose of the "single till" is to utilize the significant non-aeronautical income such as that derived from retail shops within the terminal building and rentals and parking fees to subsidize costs borne by airports in the provision of airports and air navigation services—thus reducing charges levied on aircraft operators for such services. The operative argument, at least presented implicitly, is that if not for the passengers brought by the airlines into airports, there would be no scope for operating retail shops and parking bays in the airports concerned. In other words, it is imputed that it is the airline passengers and their visitors who are the sole customers of these facilities and, therefore, the airlines which bring in these passengers should benefit from the revenue derived from such facilities. *Ex facie*, this is a sound and incontrovertible premise and it is not the purpose of this article to impugn the compelling nature of the argument. However, it is submitted that a sound legal analysis of the economic concepts underlying the modern airport is a necessity, if a thorough understanding of the subject were to be obtained. Clarification of the various positions of the airlines and airports on this issue would in turn help assess the impact of the "single till" on commercial competition of airports as autonomous business entities.

The importance of non-aeronautical revenues both for the financial viability of an airport and as one means of keeping charges on air traffic as low as possible cannot be overstated. Therefore it can be argued that as non-aeronautical revenues can only be generated at an airport because of the market created by the aircraft operators, it would be equitable that the principal users, that is the air carriers and their passengers, benefit from non-aeronautical as well as aeronautical activities. On this basis, the ICAO Secretariat suggested at ANSConf 2000 that the substance of related text in the Council Statements in Doc 9082 be reaffirmed. The Secretariat also suggested that the "single till" concept of pooling airport revenues and costs from all revenue-generating activities when charges are set apply equally in the context of air navigation services.

It was the view of Airports Council International (ACI) that using revenues from airport commercial activities to reduce aeronautical charges creates an exceptional

---

[38] Doc 9082/6 *Supra*, note 24, at p. 7.

cross-subsidy for air carriers. ACI's view was that these reduced short-term costs at the expense of longer term investment, reduced the ability and incentive of airports to develop improved commercial facilities, and in the longer term reduced commercial revenues. ACI therefore sought adoption of a more flexible policy allowing for airport operators to retain and use commercial revenues as and when needed to help fund airport improvements, in the setting of user charges and the provision of appropriate rewards to the owner of the airport, and suggested that the Conference amend Doc 9082 accordingly.

IATA was of the opinion that non-core activities within an airport perimeter, which can only develop due to aviation activities, should benefit its primary users, the air carriers, in terms of reducing the cost base for charging purposes. IATA also considered that discounts or rebates on charges were only acceptable if they comply with certain principles including, *inter alia*, that they are available to all operators and that they do not distort competition, and stressed that airports should refrain from imposing non cost-related levies on aeronautical activities directly associated with the operation of air transport services, as it considered that such levies would increase the cost of airline operations at an airport and could have discriminatory effects.

Many delegates at ANSConf 2000 supported continued endorsement of the single till concept as it reflects the special role of airports in promoting and developing air transport, serves the purpose of reducing the cost base for charges, motivates airports to develop revenues from non aeronautical activities, and provides for capital investment. The existing ICAO policy guidance was felt by some delegates to be sufficiently balanced to accommodate airports' needs. On the other hand there was considerable support for more flexibility in the interpretation and application of the single till in view of the varying situations among airports, the need to adjust to the changing airport environment, including autonomous organizational and financial structures, and investment requirements. Two delegates indicated that the existing policy in the Council Statements in Doc 9082/5 [paragraph 14(i)] does not accommodate the existing policy positions of their States.

The Conference agreed, by Recommendation 28, to submit to the ICAO Council that ICAO undertake, as a matter of high priority, a study on the application of the "single till" principle for airports as well as air navigation services, with a view to mainly identifying the elements included in it as well as determining whether any amendment is required to the Council Statements in Doc 9082 or whether there is a need for additional guidance for States. It was felt that the study should also focus on economic implications, including return on investment, and the impact on charges, take into account the need for transparency in financial accounting and flexibility to accommodate differing circumstances in different States and service environments, as well as regulatory oversight or the application of generic competition law.[39]

---

[39] Report of the Conference on the Economics of Airports and Air Navigation Services, Montreal, 19–28 June 2000, Doc 9764 ANSConf 2000, at pp. 45–46.

Today's airport, in its typical form, is primarily a commercial entity and operates as a business oriented entity. Most airports provide retail shops and parking facilities not only for airline passengers and their visitors but also to residents of the area. They are, in this sense, as much profit centres as are such retail outlets as K-Mart and Walmart. In addition, there are also airport free zones which are bonded areas, adjacent to the airport premises which, as the name suggests, are duty free areas promoting industry and other commercial activity.

The airport, like any other autonomous business enterprise, has to exploit inherent resources optionally; compete with other businesses of the area on quality of services offered; and reinvest funds in developing its business interests. The ICAO Council recognizes the continuing importance to airports of income derived from such sources as concessions, rental of premises and "free zones". The Council recommends that, with the exception of concessions that are directly associated with the operation of air transport services, such as fuel, in-flight catering and ground handling, the full development of revenues of this kind be encouraged having regard to the need for moderation in prices to the public, the requirements of passengers and the need for terminal efficiency.[40] The fact that the general public are recognized as customers of airports by ICAO policy leaves room to attribute to the wisdom of the policy statement the acknowledgment that airports do not only cater to airline passengers and their visitors, but to shoppers who may wish to pick up a bargain at the airport.

The ICAO Council Statements (Doc 9802/5) leave it wide open for airports to optimize their revenue to exceed all direct and indirect operating costs, including general administration etc., and so provide for a reasonable return on assets (before tax and cost of capital) to contribute toward necessary capital improvements.[41] This policy statement encourages commercial autonomy of airports and the practice of deviating from being instrumentalities of the State. *A fortiori*, it urges airports to optimize revenues in order to boost capital investment toward improvement, which would not necessarily mean such revenues should subsidize airport and air navigation service charges. The *Airport Economics Manual* of ICAO states:

> It should be noted that revenues from non-aeronautical activities are in fact the principal means by which a growing number of airports are able to recover their total costs because their profits from these activities more than cover the losses most of them incur on their air side operations[42] .... and in so far as ICAO cost recovery policies are concerned, not subject to the same limitations that it is recommended apply to charges on air traffic.[43]

---

[40] Doc 9082/6 *Supra*, note 24 at paragraph 24.

[41] Doc 9082/6 Id. paragraph 14 vii).

[42] *Airport Economics Manual*, First Edition—1991, Doc 9562, Chapter 6, para 6.3 at p. 50.

[43] *Id.* paragraph 4.60 at p. 40. It should also be noted that certain concession type facilities are established at an airport to provide services considered necessary for passengers, visitors and/or persons working at the airport. Such facilities may include cafeterias, post offices and tourist information counters. See *Airport Economics Manual, Supra*, note 42, at paragraph 6.11.

It is arguable in the context of the above mentioned policy that, if two different criteria are applied, one strictly on the basis of recovery of costs of services provided which would essentially be a non-profit operation, as in the case of charges levied for the provision of airport and air navigation services, and another strictly on a profit making business basis, whether the latter, applicable to non-aeronautical revenues, is mutually exclusive from the former. If this argument were to prevail, one could maintain that charges for airport and air navigation services need not as of necessity be minimized as a result of capital injection towards facilities providing such services if such capital were to be derived from the non-aeronautical activities of the airport. Of course, this does not mean that an airport which wishes to inject some capital derived from one source to another should be precluded from doing so in the best interests of aviation.

Airports, like any other business enterprises, are compelled to compete with each other to attract air traffic and business flowing therefrom and from the general public. To this end, many airports are now privatized and operate as autonomous entities. Privatization brings not only fiscal benefits, but also legal liability (as the discussion below will show) which is a possible cost factor that should be taken into account. More importantly, privatization connotes a businesslike approach that requires a certain involvement with market forces and competition that would necessitate reinvestment in the airport business for the commercial sustainability of the airport.

On the balance, however, it could be argued *a fortiori* that the advantages of airport privatization outweigh the disadvantages. For instance, privatization usually results in improved operational efficiency and aggressive market related business practice such as opening up new non traditional capital sources calculated to contribute towards financing airport infrastructure and service development. Privatization remains a part of an overall strategy to replace the public sector culture with a private sector culture by more efficient use of resources and modernization of the economy concerned.[44]

Perhaps the most important feature of privatization is that it is essentially a political process involving a change in the role of the State with the sale of State or publicly owned property, or with transfer of management from State to private sources. This inevitably leads to issues of responsibility and legal liability. For instance, an airport previously publicly owned may not be under the same rules of legal liability as it would once privatized.

The privatization process would usually involve a sustained consultation period between the parties, particularly involving the fundamental issue of the exact mode of privatization involved. Some of the options which may be considered are the creation of a new corporation whereby existing assets could be vested in the new entity and be floated publicly. Privatization could also be partial involving just some assets of the enterprise. There may also be a full public share floatation of the

---

[44] See *Report of the Ad hoc Working Group on Comparative Experiences with Privatization on its Second Session*, United Nations Conference on Trade and Development: Geneva 1993 at p. 12.

enterprise or a management buy-out structure where a company could provide backing finance in order to take the airport concerned into the private sector. There could also be a joint venture arrangement in airport privatization where private sector and government could share their equity involvements.

At the implementation stage of the privatization process, a tremendous amount of information is usually exchanged, particularly from the owners of the enterprise to the investors. Such information should demonstrate the legal rights of the parties and stipulate the rights and liabilities that would remain as residual rights and obligations of the State. A privatization process, whether it be by concession or trade sale would also entail a complex series of negotiation and contractual wrangling. Competing companies would bid against each other for the enterprise being offered for privatization.

The fundamental postulate of ICAO's policies on airport charges lies in Article 15 of the Chicago Convention which states that every airport in a contracting State which is open to public use by its national aircraft shall likewise, subject to the provisions of Article 68, be open under uniform conditions to the aircraft of all the other contracting States. The provision goes on to say that like uniform conditions shall apply to the use, by aircraft of every contracting State, of all air navigation facilities, including radio and meteorological services, which may be provided for public use for the safety and expedition of air navigation. Any charges that may be imposed or permitted to be imposed by a contracting State for the use of such airports and air navigation facilities by the aircraft of any other contracting State shall not be higher: as to aircraft not engaged in scheduled international air services, than those that would be paid by its national aircraft of the same class engaged in similar operations, and; as to aircraft engaged in scheduled international air services, than those that would be paid by its national aircraft engaged in similar international air services.

Article 15 sets out the following three basic principles:

- Uniform conditions shall apply to the use of airports and air navigation services in a Contracting State by aircraft of all other Contracting States;
- The charges imposed by a Contracting State for the use of such airports or air navigation services shall not be higher for aircraft of other Contracting States than those paid by its national aircraft engaged in similar international operations; and
- No charge shall be imposed by any Contracting State solely for the right of transit over or entry into or exit from its territory of any aircraft of a Contracting State or persons or property thereon.

While the first two of these principles do not appear to have given rise to misunderstandings, the third has, in some instances, been interpreted to mean that no charges are to be levied when an aircraft flies into, out of or over a State. That, however, is not the intent of this principle since all States are fully within their rights to recover the costs of the services they provide to aircraft operators through charges. The substance of this principle is in fact that a State should not charge solely for granting an authorization for a flight into, out of or over its territory.

6.2.4 CEANS 2008

The Conference on the Economics of Airports and Air Navigation Services (CEANS), held from 15 to 20 September 2008 in Montreal, was the latest in a periodic undertaking by ICAO and its member States to evaluate policies on the subject. The last time such an event took place was in June 2000 when member States gathered at the Conference on the Economics of Airports and Air Navigation Services (ANSConf 2000).[45]

CEANS was preceded by a symposium on 14 September 2008 convened by ICAO to discuss and introduce the participants to the various issues to be discussed at the Conference and had three Agenda Items: issues involving interaction between States, providers and users; specific issues related to airport economics and management; and specific issues related to air navigation services economics and management. These items were in turn divided into several sub items.

Mr. Roberto Kobeh Gonzalez, President of the ICAO Council said in his opening address at CEANS that the conference was the seventh in a series since the first conference on charges in 1956. He pointed out that that CEANS was timely in that significant developments of the past decade on airports and air navigation services had necessitated a review of the organization and financing of service providers. Mr. Gonzalez went on to say that whereas ANSConf 2000 had updated ICAO's policy guidance in the areas of commercialization and privatization of airports and air navigation service providers, CEANS was intended to build on recent experiences in these areas and further develop policy guidance with a view to enabling States to address effectively the regulatory issues pertaining to charges and economic aspects of airports and air navigation services.

Mr. Saud A.R. Hashem, Chairman of the Air Transport Committee of the Council of ICAO stated in his address that CEANS was expected to bring for review a number of new trends in charging practices such as the allocation of cost on a per passenger basis; the application of differential charges with respect to introductory discounts for new services and low cost carriers; and access to airport facilities in terms of potential unfair treatment or discrimination. Mr. Hashem also stressed the importance of evaluating at CEANS the risk inherent in low awareness within commercialized and privatized airports and air navigation service providers with regard to ICAO's policies and guidance material in the economic field.

The Conference, which was attended by 103 member States of ICAO and 17 observer organizations met as a single body and held 10 meetings. There were

---

[45] See Report of the conference on the economics of airports and air navigation services: air transport infrastructure for the twenty-first century. Montreal, 19–28 June 2000. Doc 9764, ANSConf 2000. ICAO: Montreal, 2000. For a discussion on ANSConf 2000 see Abeyratne (2001), pp. 217–230.

no summary minutes.[46] However, a full Report[47] of the Conference has been issued by ICAO.

6.2.4.1   Discussions at CEANS

The Conference discussed the following issues pertaining to the interaction between States, airports and air navigation service providers.

(a) **Economic oversight**: CEANS recognized that the protection of users against potential abuse of dominant position by airports and air navigation services providers (ANSPs) is a State's responsibility and that this responsibility could be discharged through economic oversight, i.e. monitoring by a State of the commercial and operational practices of service providers. The Conference proceeded to discuss the scope and forms of economic oversight, as well as the associated costs and benefits, and considered the various forms of economic oversight that might be applied, in accordance with the specific circumstances in a State;

(b) **Economic performance and minimum reporting requirements**: It was acknowledged that the assessment of airport and ANSP performance, including the use of benchmarking, can serve to improve safety, quality of services, productivity and cost-efficiency of airports and ANSPs, as well as to support investment decisions. The Conference discussed the kind of data that would be fundamental to performance reporting, as well as the minimum reporting requirements for performance measurement, with a view to achieving a mutual understanding by all stakeholders.

(c) **Consultation with users**: Discussions revolved round the need for good relations between regulators, providers and users which is important for the effective development of air transport. However, it was noted that many airports and ANSPs around the world either do not consult users, or maintain a proper and regular consultation process. The Conference reviewed existing consultation mechanisms between airports/ANSPs and users, and considered innovative solutions to establish the foundation for a sound cooperation between providers and users.

With regard to specific issues related to airport economics and management the Conference discussed the following:

(a) **Governance, ownership and control**: CEANS was of the view that changes in governance, ownership and control of airports, including cross-border

---

[46] For purposes of CEANS, the Council of ICAO suspended Rule 26 of the Standing Rules of Procedure for Meetings in the Air Transport Field (Doc 8683) which calls for summary minutes of meetings convened by the ICAO Council.

[47] Report of the Conference on the Economics of Airports and Air Navigation Services (CEANS) Montreal, 15–20 September 2008, Doc 9908;CEANS 2008.

investments in privatized airports, can have implications for a State's obligations in the provision and operation of airport services. Based on the experiences of commercialization and privatization, the Conference considered the influence of different governance structures on the performance of airports and what measures States can take in order to ensure that all relevant obligations of States are observed.

(b) **Cost basis for charges**: On the basis that the cost basis of an airport for charging purposes has usually been established by taking into account the costs of operation and maintenance, cost of capital and depreciation of assets (based on historical value in most cases), and a "reasonable" return on assets, CEANS approached the issue by attempting to build a consensus on possible ways to assess what would constitute a "reasonable" rate of return, and explore the possibility of consolidating several airport cost bases into one cost base for charging purposes.

(c) **Cost allocation and charging systems**: It was recognized that ICAO's current policies and guidance material on airport charges have provisions dealing with how the costs of the various airport facilities and services should be allocated to different categories of users. However, the Conference took note of the fact that some new trends have emerged such as the allocation of costs on a per passenger basis, which includes all or most cost bases of the aeronautical activities. The Conference's discussions revolved round the issue as to whether such new approaches are consistent with ICAO's policies, and consideration was given to appropriate amendments to the policies and guidance material on cost allocation.

(d) **Non-discrimination aspects**: On the subject of non-discrimination, CEANS recognized that, in recent years, airport operators have developed certain differential charges to attract and retain new airline services, for example, discounts on passenger service charges and incentive schemes for particular airlines, including low-cost carriers. It was also noted that some of these differential charges might be non-transparent, discriminatory and anti-competitive, especially when they constitute a form of State aid. The Conference addressed the issue as to how to deal with the measures taken by airport operators that have the potential to create unfair treatment, as well as the issue of access to airport facilities.

(e) **Financing and cost recovery of security measures**: According to ICAO's policies on security charges, the costs of security functions performed by States such as general policing, intelligence gathering and national security should not be passed on to the airport users. However, it was noted that practices differ between regions and States, which have financial implications on users. The Conference reviewed the current policies and discuss how to achieve a more harmonized implementation of ICAO's policies regarding airport security charges.

On the subject of specific issues related to air navigation services economics and management, the following views were exchanged:

(a) **Governance, ownership and control**: CEANS noted that, while autonomous and commercialized ANSPs have been established in many States around the World, financing was still an issue in other States. Therefore the Conference considered the importance of further promotion of ICAO's policies on the establishment of autonomous ANSPs, including separation between regulatory and operational functions. The Conference also reviewed commercialization experiences and discussed their influence on the performance of ANSPs, with particular attention to governance and management structures.

(b) **Cost basis for charges**: On the basis that the cost basis of an ANSP for charging purposes has usually been established by taking into account the costs of operation and maintenance, plus cost of capital and depreciation of assets, as well as a "reasonable" return on assets, the Conference endeavoured to build a consensus on possible ways to assess what would constitute a "reasonable" rate of return for ANSPs. Another issue that the Conference addressed was the practice of contingency funds in order to cater for unforeseen severe drops in traffic/revenues.

(c) **Cost allocation and charging systems**: The Conference reviewed recent developments, and considered the need for and implications of an alternative categorization of services based on the portions of airspace and on the phase of flight, which could lead to the introduction of zone (differential) charges. The Conference also discussed the relevance of the element of aircraft weight in the charging formulae and incentives that could apply both to providers and users. In its deliberations, CEANS based its observation on the fundamental fact that the allocation of costs to air traffic control services has traditionally followed the categorization of services between aerodrome control, approach control and area control.

(d) **Economic and organizational aspects related to implementation of the global air traffic management (ATM) concept**: It is an established fact that the global air traffic management (ATM) concept seeks to derive operational, economic as well as environmental benefits for all members of the ATM community. The efficient and cost effective implementation of a seamless ATM system will be facilitated through multilateral cooperation and institutional arrangements for financing and charging. Taking these considerations into account, CEANS reviewed the work of ICAO concerning the economic, organizational and managerial aspects of the implementation of the global ATM concept, and considered what practical guidance and support ICAO may provide to States, as well as to regional and sub-regional entities, including guidance on charging systems.

6.2.4.2 Recommendations of CEANS

Using its deliberations as a base, **CEANS** adopted several recommendations to be submitted for the approval of the Council of ICAO. On **economic oversight** CEANS recommended that, since the responsibility for economic oversight devolves upon each State, and in order to preclude possibilities of abuse of dominant position by the service provider, States should ensure that service providers consult with users and that appropriate performance management systems are developed and implemented by their service providers. The recommendation went on to state that States should select the appropriate form of economic oversight according to their specific circumstances, while keeping regulatory interventions at a minimum and as required. It was suggested that, when deciding on an appropriate form of economic oversight, the degree of competition, the costs and benefits related to alternative forms of oversight, as well as the legal, institutional and governance frameworks should be taken into consideration. It was also recommended that States should consider adoption of a regional approach to economic oversight where individual States lack the capacity to adequately perform economic oversight functions. Finally, this recommendation states that in order to achieve the objectives of economic oversight and to assist States in this regard, ICAO should amend *ICAO's Policies on Charges for Airports and Air Navigation Services* contained in Doc 9082 to clarify the purpose and scope of economic oversight for airports and air navigation services with reference to its different forms and the selection of the most appropriate form of oversight.

On the subject of **economic performance and minimum service requirements** the CEANS recommendation was that States should ensure that their service providers establish performance objectives with the purpose, as a minimum, to continuously improve performance in four key performance areas (KPAs), i.e. safety, quality of service, productivity and cost-effectiveness, and to report at least one relevant performance indicator for each KPA. Towards achieving this objective, States may choose additional KPAs according to their objectives and their particular circumstances. This recommendation also suggested that ICAO should amend Doc 9082 to recommend the establishment of performance management systems by service providers and to include the major elements of a performance management system with emphasis on the selection of KPAs and related indicators. Finally it was recommended that ICAO establish a dialogue, where appropriate, with regional organizations on economic performance with a view to improving performance of the air navigation services system.

On **consultation with users**, CEANS recommended that States should ensure, within their economic oversight responsibilities that a clearly defined, regular consultation process is established with users by their airports and air navigation services entities where provider/user cooperative arrangements are not already in place. It was further recommended that States should ensure that users are consulted on the level and structure of charges as well as on capacity development and investments; Another duty that devolves on States in this regard is to consider users' feedback obtained during consultations as far as possible before reaching a decision regarding any proposal; that the confidentiality of the market-sensitive data is properly protected; and that the relevant decision documents provide appropriate rationale for the decision. This recommendation also included a clause to the effect that ICAO amend Doc 9082 to include the concept of a clearly defined, regular consultation process under new paragraphs on consultation with users, covering both airports and air navigation services.

On **governance, ownership and control of airports** the recommendation was that States should consider the establishment of autonomous entities to operate airports, taking into account the economic viability of the airport as well as the interests of service providers and users. CEANS also recommended that where the operation of one or more airports represents only one of several functions performed by a government entity, States should give consideration to a clear separation of the regulatory and operational functions, with roles and powers clearly defined for each one.

States are called upon by virtue of this recommendation to review the governance structure with regard to their airports, and ensure the use of best practices of good corporate governance with regard to objectives and responsibilities, shareholders' rights and their treatment, responsibilities of the board, power and accountability of the management, relationship with interested parties, and information disclosure. Furthermore, it was suggested that whenever an autonomous entity is established, States should ensure that all relevant obligations of the State under the Chicago Convention, its Annexes and in air services agreements are complied with and that ICAO's policies and practices are observed. Again, ICAO was called upon to amend Doc 9082 to emphasize the importance of separation of the regulatory and operational functions and to include the components needed to ensure good governance through the application of best practices. Another recommendation pertaining to ICAO on this subject was that the Organization should consider developing its policies and guidance material for the management and operation of airports which are not economically viable but are necessary, as part of an integrated transport network, for the safety and security of international air transport, as well as for socio-economic purposes. An example cited for ICAO's consideration is the work carried out by the Latin American Civil Aviation Commission (LACAC) on airport concession services.

On **aggregation of cost bases** CEANS recommended that ICAO should amend Doc 9082 with a view to allowing more flexibility in setting airport charges. This calls for airports to maintain cost data in adequate detail which ensures transparency and oversight and the avoidance of discrimination in setting charges.

On **rate of returns for airports and air navigation services** it was recommended that within their economic oversight responsibilities, States should, where necessary and in the light of national circumstances, clearly define the methodology for determining what is a reasonable rate of return on assets for their service providers. ICAO was called upon to develop additional guidance material regarding possible methodologies to assess the risk element and the value of assets in the context of the determination of a reasonable rate of return.

On **differential charges** CEANS has recommended that within their economic oversight responsibilities States should, where necessary, assess the positive and negative effects associated with specific forms of differential charges applied by airports on a case-by-case basis according to national circumstances. Furthermore, States were called upon to ensure that differential charges are offered on a non-discriminatory basis; that they are transparent in terms of their creation, purpose and the criteria on which they are offered; that, without prejudice to modulated charging schemes, costs associated with differential charges are not allocated, either directly or indirectly, to those other users not benefiting from them; and that, if the purpose is to attract and/or retain new air services, they are offered only on a temporary basis. ICAO was called upon to amend Doc 9082 to reflect the principles of transparency and time limitation for start-up aids in the application of differential charges.

Access to airport infrastructure was another important issue under discussion. Here, the recommendation was that States should give due consideration to the results of ICAO's studies and relevant guidance on slot allocation, as well as international general and business aviation access to airports, at their discretion and in a flexible manner. ICAO was again called upon to amend Doc 9082 to give more emphasis to the importance of international general aviation by including a special reference to business aviation. Furthermore, the ICAO Council was called upon to take appropriate action on the issues related to slot allocation and night curfews.

On **cost recovery of security measures at airports** it was recommended that since the current ICAO policies for the cost recovery of security measures at airports are still adequate, States should ensure their implementation in order to foster harmonization worldwide.

Governance, ownership and control of air navigation service providers was another subject where a recommendation was made. CEANS recommended that where States choose not to establish autonomous providers of air navigation services, they should give consideration to a clear separation of the regulatory and operational functions, with roles and powers clearly defined for each one. Furthermore it was recommended that States should review the governance structure with regard to their ANSPs and ensure the use of best practices of good corporate governance with regard to objectives and responsibilities, shareholders' rights and their treatment, responsibilities of the board, power and accountability of the management, relationship with interested parties, and information disclosure. This was another area where ICAO intervention was sought in reviewing Doc 9082 with a view to emphasizing the importance of the separation of regulatory and operational functions and to include the components needed to ensure good governance through the application of best practices.

On **categorizing services for cost allocation purposes** it was recommended that States might consider the use of the alternative categorization of services and the corresponding zone charges, depending on their particular circumstances, for example in congested or complex airspace, in order to achieve a more cost-related charging system and an efficient use of resources. ICAO was requested to expand its guidance material on cost allocation between portions of airspace and/or phases of flight, as well as develop new guidance material on zone charges. There was also a brief recommendation that ICAO should review its guidance material on aircraft weight in charging formulae and adjust it as necessary.

The Conference also discussed the issue of **incentives that could apply to service providers** and encouraged States to introduce appropriate forms of incentives for ANSPs, within their economic oversight responsibilities, with a view to optimizing the use and delivery of air navigation services, reducing the overall cost of such services, and increasing their efficiency, all within ICAO guidelines already in place in Doc 9082. On the implementation and operation of ICAO's Global Air Navigation Plan (GANP) it was recommended that ICAO amend Doc 9082 to emphasize the need for international cooperation, as well as to refer to regional approaches, in the implementation of the global ATM operational concept and the GANP. States were encouraged to facilitate and strive for the efficient and cost-effective implementation of the global ATM operational concept, using the Global Air Navigation Plan (GANP) as the implementation planning document, through international cooperation and collaboration within the ATM community. ICAO was requested amend Doc 9082 so that the need for international cooperation could be emphasized, as well as to refer to regional approaches, in the implementation of the global ATM operational concept and the GANP.

The last subject taken up by CEANS for discussion was the implementation of ICAO's policies on charges. Here the recommendation was that States recognize that adherence to ICAO's policies on charges in Doc 9082, which have their principal origin in Article 15 of the Chicago Convention, promotes the efficient and cost-effective provision and operation of airports and air navigation services, as well as a sound relationship with users, in particular, with respect to transparency and the fair treatment of different categories of users and that States ensure that their airports and ANSPs adhere to ICAO's policies on charges. It was also recommended that ICAO should encourage States to adopt the principles of Doc 9082 on non-discrimination, cost-relatedness, transparency and consultation with users into national legislation, regulation or policies to ensure compliance by airports and ANSPs. Finally, ICAO was called upon to encourage States to incorporate the principles of Doc 9082 on non-discrimination, cost-relatedness, transparency and consultation with users into their future air services agreements to ensure compliance by airports and ANSPs.

The main purpose of CEANS was for ICAO member States to discuss revisions to existing policy and guidance material on the economics of airports and air navigation services and to request the Council to incorporate such revisions in Doc 9082. The ICAO Council, at its eleventh meeting of the 185th Session on 14 November 2008, took up for consideration the CEANS Report and approved all the recommendations contained in the Report. Accordingly, the ICAO Secretariat will proceed to produce a new version of Doc 9082.

It is expected that these new revisions would assist States in ensuring improved performance of airports and air navigation service providers and ensure transparency and consultation in the implementation of ICAO policy in their territories.

6.2.5   ICAO's Policies on Airport Charges

As discussed earlier, income of airports are derived from aeronautical and non aeronautical charges imposed by airports on their primary client—the airline which operates into and out of the airport, and from rentals of premises. The vexed issue of the need to reach consensus on a just and equitable basis for the imposition of charges levied on airlines by airports for services rendered has been the subject of discussion at many ICAO conferences. Part of the problem has been that airports have, over the years, been privatized and commercialized, necessitating them to be operated in a businesslike manner. However, some core issues have remained unchanged, the first being that, after everything is said and done, there is only one product in the air transport business and that is air transport which is provided by the airlines. The second is that it is an immutable principle that the State is ultimately responsible for meeting the needs of the people of the world for safe, regular, economic and efficient air transport services, working through ICAO, as per the Chicago Convention. The blurring of concepts that has arisen in meshing these fundamental principles brings to bear the need to critically appraise one area that exemplifies the confusion—airports charges for services provided to airlines. This article critically appraises the issue and identifies certain anomalies that exist.

At the very core of the rationale for charging airlines for services rendered to them by airports is the Universal Declaration of Human Rights of the United Nations.[48] Article 7 of the Declaration states that all are equal before the law and are entitled without any discrimination to equal protection of the law. The provision goes on to say that all are entitled to equal protection against any discrimination in violation of the Declaration and against any incitement to such discrimination. Article 17 provides that everyone has the right to own property alone as well as in association with others[49] and that no one shall be arbitrarily deprived of his property.[50] Since property includes money[51] and charges levied on airlines by airports comprise money, the Declaration could be legally construed as prohibiting arbitrary charging in excess of amounts that correspond to the services rendered.

It has to be noted that the Declaration is not a treaty and therefore not a binding source of formal law. Therefore it is not a self-executing document and persons relying on the enforcement of these principles would have to rely on the justiciability of treaties that implement the Declaration.[52] However, the Declaration

---

[48] Adopted and proclaimed by General Assembly Resolution 217 A (III) of 10 December 1948.

[49] Article 17(1).

[50] Article 17(2).

[51] See The Public International Law of Taxation, Text, Cases and Materials (Asif Hasan Qureshi ed.) Graham & Trotman Ltd:1994 at 295.

[52] For example, on 4 November 1950 the Council of Europe member States signed the Convention for the Protection of Human Rights and Fundamental Freedoms, also known as the Rome Convention, which implemented the principles of the Universal Declaration of Human Rights and, in Article 25 accorded an individual the right to complain if his rights enshrined in the Declaration were eroded.

remains a statement of moral principles that is calculated to have a coercive influence on the community of nations. With this in mind, the starting point, as the moral denominator for charging, would be to recognize that to impose charges which are not commensurate with the services rendered would be tantamount to the levy of a tax that unjustly enriches the airport and the State concerned. In this context, it is worthy of note that ICAO, for the purpose of its policy objectives, makes a distinction between a charge and a tax, in that charges are levies to defray the costs of providing facilities and services for civil aviation while taxes are levies to raise general national and local government revenues that are applied for non-aviation purposes.[53]

Specific regulatory provisions applicable to charges levied by airports have their genesis in ICAO and are contained in Doc 9082[54] which, in the Foreword has the curious opening: "ICAO's policies on charges for airports and air navigation services which follow contain the recommendations and conclusions of the Council". It must be mentioned that the earlier version of this document used the words "statements" instead of "recommendations and conclusions". The former is seemingly more appropriate since the Chicago Convention does not empower the Council to arrive at recommendations and conclusions either in Article 54—which contain the mandatory functions of the Council; and Article 55—which lays down the permissive functions of the Council. These two provisions do not require the Council to issue guidance to States on any matter pertaining to air transport. However, Article 54 (b) makes it a mandatory function of the Council to carry out the directions of the Assembly, although there is no clear definition of the word "directions". If this word were to be interpreted to include a request of the Assembly, one could apply resolving clause 5 of Assembly Resolution A36-15[55] which requests the Council to ensure that the guidance and advice (not conclusions and recommendations) contained in Doc 9082 are current and responsive to the requirements of Contracting States. In this context, one can only laud the ICAO Council for taking a leadership role and for taking the initiative to publish its recommendations and conclusions, however inconsistent the words may be between the Assembly Resolution and those used in Doc 9082.

The inconsistency of wording does not stop there. Paragraph 8 (i) of Doc 9082 states that the Council recommends that States permit the imposition of charges only for services and functions which are provided for, directly related to or ultimately beneficial for, civil aviation operations. The anomaly lies in the word "functions" which is not defined or elaborated anywhere in the document. Does this give airports the licence to levy charges on airlines for "functions" as defined or

---

[53] *ICAO's Policies on Taxation in the Field of Air Transport*, Doc 8632, Third Edition: 2000, at 3.

[54] In Chap. 7. *ICAO's Policies on Charges for Airports and Air Navigation Services* Doc 9082/9 Ninth Edition-2012.

[55] Resolution A36-15, Consolidated Statement of Continuing ICAO Policies in the Air Transport Field, *Assembly Resolutions in Force* (as of 28 September 2007) Doc. 9902, III-1 at III-13.

determined by them? Or, could one take the wording of Doc 8632[56] and apply the word "facilities" which is used therein as being meant by the word "functions" in this context?

Recent trends in regulatory control of the levy of airports charges dictate that a close look should be taken with view to evaluating whether there exists an environment for the imposition on airlines of airports charges on a just and equitable basis. This article will discuss this issue with emphasis on currently applicable regulatory provisions. Although the ensuing discussions will focus only on airports charges, it must be noted that the applicable ICAO regulatory policy also applies *ex aequo* to air navigation services.

6.2.5.1   Current Regulatory Provisions

The current policies of the ICAO on charges for airports and air navigation services stemmed from the recommendations of the Conference on the Economics of Airports and Air Navigation Services (ANSConf 2000) which were endorsed by the Council of ICAO.[57] ANSConf 2000, which was held in Montreal on 19–28 June 2000, came to the conclusion that the profile of basic cost recovery policy may need to be raised.[58] It was recommended by the Conference that this measure could be adopted within the parameters of existing policy calling for revenues from charges levied on international civil aviation and it would only be applied towards defraying the costs of facilities and services provided for international civil aviation. It was also recommended that revenues from other sources than charges on air traffic shall be taken into account before the cost basis for charges on air traffic are determined. ICAO advised the Conference that airports and air navigation services may produce sufficient revenues to exceed all operating costs and so provide for a reasonable return on assets to contribute towards necessary capital improvements. Of course, the governing principle would be that consultation with users shall take place before significant changes in charging systems or levels of charges are introduced.[59]

---

[56] *Supra*, note 53.

[57] See Report of the conference on the economics of airports and air navigation services: air transport infrastructure for the twenty-first century. Montreal, 19–28 June 2000. Doc 9764, ANSConf 2000. ICAO: Montreal, 2000. For a discussion on ANSConf 2000 see Abeyratne (2001), pp. 217–230.

[58] *ANSConf-WP/4* at para. 5.1.

[59] *Id*. para. 5.3. ICAO's recommendations to ANSConf 2000 were both timely and practical, given the evolving fabric of economic forces which now govern airports and air navigation services. The recommendations also stimulate some reflection on the complexities of financing principles now applicable to the services provided by airports and air navigation services providers. In substance, the issue of costing and pricing of services would be dependent upon underlying practices and economic factors as the bunching of aviation and non-aviation revenues and their effect on the overall pricing policy relating to airports and air navigation services and a significant paradigm shift from Article 15 of the Chicago Convention.

The baseline of ICAO's policies on charges lies in Article 15 of the Chicago Convention, the basic philosophy of which is that every airport in a Contracting State which is open to public use by its national aircraft shall likewise be open under uniform conditions to the aircraft of all the other Contracting States. It also requires that uniform conditions shall apply to the use, by aircraft of every Contracting State, of all air navigation facilities, including radio and meteorological services,[60] which may be provided for public use for the safety and expedition of air navigation.[61] Article 15 subsumes three fundamental postulates:

(a) Uniform conditions should apply in the use of facilities provided by airports and air navigation services;
(b) Aircraft operators should be charged on a non-discriminatory basis; and
(c) No charges should be levied for the mere transit over, entry into or exit from the territory of a Contracting State.

Current ICAO policy also recognizes that the financial situation of airports and air navigation services are in a constant state of evolution and that the financial situation of the primary users, the scheduled airlines, generally fluctuates with the performance of national, regional and global economies.[62] Accordingly, the ICAO Council recommends that States permit the imposition of charges only for services and functions which are provided for, directly related to, or ultimately beneficial for, civil aviation operations. States are therefore encouraged to refrain from imposing charges which discriminate against international civil aviation in relation to other modes of transport.[63]

ICAO's policies are at best only authoritative in practice and, from a legal perspective, are rendered destitute of effect by the acknowledged lack of enforcement power afflicting them. In this context it is curious that, six decades after the establishment of ICAO some still refer to its powers and functions.[64] There are some others who allude to ICAO's mandate. The fact is that ICAO has only aims and objectives, recognized by the Chicago Convention[65] which established

---

[60] Article 28 of the Chicago Convention calls on each Contracting State, so far as it may find practicable, to provide airport and air navigation facilities, in accordance with the standards and practices recommended or established in pursuance of the Convention.

[61] Article 15 also provides that any charges that may be imposed or permitted to be imposed by a Contracting State for the use of such airports and air navigation facilities by the aircraft of any other Contracting State shall not be higher: as to aircraft not engaged in scheduled international air services, than those that would be paid by its national aircraft of the same class engaged in similar operations; and as to aircraft engaged in scheduled international air services, than those that would be paid by its national aircraft engaged in similar international air services.

[62] *ICAO's Policies on Charges for Airports and Air Navigation Services, supra*, note 54, paragraph 7.

[63] *Id.* Paragraph 8. Paragraph 9 that follows states that the Council is concerned over the proliferation of charges on air traffic and notes that the imposition of charges in one jurisdiction can lead to the introduction of charges in another jurisdiction.

[64] MacKenzie (2008), Preface at 1.

[65] *Supra*, note 1 in Chap. 1.

the Organization.[66] Broadly, those aims and objectives are to develop the principles and techniques of international air navigation and to foster the planning and development of international air transport. In effect, this bifurcation implicitly reflects the agreement of the international community of States which signed the Chicago Convention that ICAO could adopt Standards in the technical fields of air navigation and could only offer guidelines in the economic field.

In its basic documentation, the ICAO Council notes that with the rapidly growing autonomy in the provision of airports and air navigation services, many States may wish to establish an independent mechanism for the economic regulation of airports and air navigation services[67] To this end the Council recommends *inter alia* that States should ensure there is no overcharging or other anti-competitive practice or abuse of dominant position.[68] The Council further states that for the successful collection of charges for airports and air navigation services entities, it is essential that a collection policy be established by an airport or air navigation services entity, or where applicable by a State.[69] In this regard the cost basis for airport charges is an important issue and the Council considers that, as a general principle it is desirable, where an airport is provided for international use, that the users shall ultimately bear their full and final share of the cost of providing the airport. It is therefore considered important that airports maintain accounts which provide information adequate for the needs of both airports and users and that the facilities and services related to airport charges be identified as precisely as possible.[70] It is interesting, once again to note that the term "facilities" has been used and that there is no mention of the word "functions". The cost to be shared is the full cost of providing the airport and its essential ancillary services, including appropriate amounts for cost of capital and depreciation of assets, as well as the cost of maintenance, operation, management and administration. However, there is a caveat that the costs to be shared must allow for all aeronautical revenues plus contributions from non-aeronautical revenues accruing from the operation of the airport to its operators.[71] The Council also states that the proportion of costs allocable to various categories of users, including State aircraft, should be determined on an equitable basis, so that no users shall be burdened with costs which are not properly allocable to them according to sound economic principles.

The aforesaid provisions have the underlying requirement of economic oversight if the regulators were to ensure that charges are being levied in a just and equitable manner.

---

[66] *Id.* Article 43. This article provides that an organization to be named the International Civil Aviation Organization is formed by the Convention. It is made of an Assembly, a Council, and such other bodies as may be necessary.

[67] Doc 9082, *supra*, note 54 at paragraph 15 (ii).

[68] *Id.* paragraph 15 (ii).

[69] *Id.* paragraph 18.

[70] Doc 9082, *op. cit.*, note 8 paragraph 21.

[71] *Id.* paragraph 22 (i).

6.2.5.2   The Legal Status of ICAO Policy

Although there was much discussion at CEANS on "giving teeth" to ICAO policy in order to ensure economic oversight by States of their airports and air navigation services providers, the Conference failed to arrive at a consensus on including text in the recommendations to the effect that States should incorporate the principles enunciated in ICAO document 9082 in their legislation or rules. The end result was a somewhat watered down recommendation that States should select the appropriate form of economic oversight according to their specific circumstances, while keeping regulatory interventions at a minimum and as required. Most delegations were, quite rightly, reluctant to agree to a recommendation that would impose upon States an obligation to incorporate policy guidelines into national legislation. Furthermore, the Conference correctly noted that States differed considerably in their economic circumstances and demand for services rendered by airports and air navigation services in their territories and therefore should be left to decide the best course of economic oversight to be taken in their territories.

ICAO's economic policies emanate from the States. However, these policies are no more than consensual principles that offer policy guidance and are at best left to the discretion of the States to follow. To require or recommend that States incorporate such policy in their national legislation or regulations is a reversal of the empowerment of ICAO by States upon which ICAO is founded, whereby ICAO is enabled by States to pursue its aims and objectives under Article 44 of the Chicago Convention.[72]

No international body or institution can legitimately expect a sovereign State to incorporate, as national legislation, policies that the former adopts. In this case, such a policy directive would come from the ICAO Council, the mandatory functions of which are stipulated in Article 54 of the Chicago Convention.[73] Nowhere in either the mandatory or permissive functions (contained in Article 55) is the Council given authority to act as legislator or regulator.[74] Even if such a function were to be elevated to the level of the ICAO Assembly, a resolution of the Assembly cannot require or even recommend that its principles be incorporated into national law or regulation. *Brownlie* has expressed the view that decisions by international conferences and organizations can in principle only bind those States

---

[72] The overarching aims and objectives of ICAO, as contained in Article 44 of the Convention is to develop the principles and techniques of international air navigation and to foster the planning and development of international air transport so as to meet the needs of the peoples for safe, regular, efficient and economical air transport.

[73] The closest the Council comes in this respect is to adopt morally binding Standards and Recommended Practices, where, as per Article 54 (I) of the Chicago Convention the Council may adopt international standards and recommended practices and for convenience, designate them as Annexes to the Convention and notify all Contracting States of the action taken.

[74] For a discussion on the role of the Council in this context see Abeyratne (1992), pp. 387–394. See also Abeyratne (1997), pp. 395–412.

accepting them.[75] Shaw, referring to the binding force of United Nations General; Assembly Resolutions states:

> ...one must be alive to the dangers in ascribing legal value to everything that emanates from the Assembly. Resolutions are often the results of political compromises and arrangements and, comprehended in that sense, never intended to constitute binding norms. Great care must be taken in moving from a plethora of practice to the identification of legal norms.[76]

With regard to the practice of other international organizations, a little more caution might be required, as a resolution might create a custom. Non-binding instruments form a special category that is sometimes referred to as "soft law" which is definitely not law in the sense of enforceability.[77]

The above discussion brings to bear some anomalies that exist in the field of airports charges. The first is that, according to ICAO policy, four elements are critical for prudent charges policy: non-discrimination (as enshrined in Article 15 of the Chicago Convention); transparency; cost relatedness; and consultation. The first two elements are self-explanatory. However, cost relatedness and consultation are open to interpretation. One could argue that cost related charges need not necessarily be restricted to actual costs but could be geared to earn profits for the airports as long as such costs are calculated in relation to the cost of services provided. With regard to consultation, there have been instances where the service provider has met with users and other stakeholders and merely informed them that certain charges were to be increased.

The second anomaly is that, the statement in Doc 9082—that autonomy and privatization of airports are preferred modes of operating airports[78]—has inadvertently resulted in the obfuscation of the fundamental principle that the ultimate responsibility for the setting and levying of charges rests with the State concerned. Although by and large States have been observed to follow ICAO policy in this area, there are many airports today which set charges and impose them, with a cursory and *pro forma* notice to the State concerned which invariably approves it. The Chicago Conference of 1944, which resulted in the adoption of the Chicago Convention, in its consideration of draft Article 15 of the Convention at that time, has explicitly recorded that "[E]ach contracting State shall establish scales of charges for the use of such airports and air navigation facilities which shall be uniformly applicable to the aircraft of all other States..."[79] Article 15 of the Chicago Convention follows this approach when it states: "Any charges that may be imposed or permitted to be imposed by a contracting State..."clearly implying that it is the State which is responsible for the imposition of charges.

---

[75] Brownlie (1990), p. 691.

[76] Shaw (2003), p. 110.

[77] *Id.* 111. See also Tammes (1958) at 265.

[78] Doc 9082 *supra*, note 54 at paragraph 10.

[79] *Proceedings of the International Civil Aviation Conference*, Chicago: Illinois November 1 to December 7, 1944, Vol. 1, at 663.

The third anomaly is that Doc 9082 which sets policy at a high level, is open to interpretation as some of its key provisions, as pointed out in the Introduction to this article, may open the possibility for interpretation and subjective treatment of critical principles concerned with ICAO's charging policy. This notwithstanding, there is no room for doubt that Doc 9082 is a generally clear policy statement which has served ICAO contracting States well and provided guidance over the years.

This discussion brings one to the conclusion that, ultimately, the responsibility clearly lies with the States, which not only have to oversee airport charges but to provide the necessary economic oversight to ensure that charges are levied justly, equitably and in a prudent manner. One effective way of ensuring this is for States to include the four elements of transparency; non-discrimination; cost relatedness and consultation in their bilateral air services agreements.

6.3 Economic Regulation and Security of Airlines: The Importance of Cargo

Airlines have to conform to the provisions of the Chicago Convention, even though much of the responsibilities devolve upon States themselves under the Convention. Article 86 of the Convention requires that, unless the Council decides otherwise any decision by the Council on whether an international airline is operating in conformity with the provisions of this Convention shall remain in effect unless reversed on appeal. On any other matter, decisions of the Council shall, if appealed from, be suspended until the appeal is decided. The decisions of the Permanent Court of International Justice and of an arbitral tribunal shall be final and binding. Article 87 prescribes penalties for non-conformity of an airline with the Convention when it says that each contracting State (to the Convention) undertakes not to allow the operation of an airline of a contracting State through the airspace above its territory if the Council has decided that the airline concerned is not conforming to a final decision rendered in accordance with the previous Article (Article 86).

There has already been a discussion on the liberalization of air transport in the context of the passenger. The other aspect of economic regulation lies in the carriage of cargo. At ATConf/6 this subject came under close scrutiny, along with the aspect of the inter-relationship with liberalization and security. The Conference noted that air cargo services have become increasingly important in economic development and world trade. According to corroborating sources, in 2011, the total goods carried worldwide by air represent about 2 % of global trade by volume, but around 40 % by value. A majority of high-value goods rely on transport by air. Forecasts suggested that over the next 20 years, the global air cargo market will expand at an annual rate of 5.2 %, reflecting increased trade through liberalization of markets.

The Conference noted that, in the past two decades, international air cargo services had benefitted from an increasing number of liberalized agreements. For example, as at the end of October 2012, of the 400 plus open skies agreements concluded by States, more than 100 granted Seventh freedom for air cargo or all cargo services, thus providing greater opportunity for the growth of such services. ICAO submitted to the Conference that expanding freedom for air cargo services has been a subject of examination by ICAO, States and the industry in the last two air transport conferences in the context of liberalizing market access. ICAO has also developed related guidance and regulatory arrangement, such as the bilateral model clause (an annex on air cargo) endorsed by ATConf/5, which is included in the ICAO Template Air Services Agreement (TASA) for use by States.

In terms of possible approaches for liberalizing air cargo services, ATConf/5 recommended that States may do so at their discretion unilaterally, bilaterally or through a multilateral approach. This guidance continues to be relevant today. It was ICAO's suggestion that, as agreed at ATConf/5, air cargo and, in particular, all cargo operations should be considered for accelerated liberalization. In this regard, both States and ICAO can play a role in advancing liberalization. States can, within the existing air service agreement framework, consider granting more freedom and operational flexibility for air cargo services, catering to distinct features and requirements of air cargo operations and customers.

ATConf/6 noted that ICAO can play a leadership role in exploring multilateral approaches, as recommended by ATConf/5. One approach ICAO could consider is the development of a specific international agreement to facilitate further liberalization of all cargo service. Such an agreement could be built on existing provisions, such as the ICAO model clause on air cargo or those included in open skies agreements to provide traffic rights and operational flexibility.

Such an undertaking would require careful study and preparation, including consultation with experts, States, and aviation stakeholders. In light of the different conditions and needs of States, the agreement could be prepared for signature for willing and ready parties initially, and opened for subsequent accession by others. Such an action, if agreed by the international aviation community, would represent a significant step forward in the development of sustainable international air transport. It should be noted that a recent survey of States in October 2012 revealed that 78 % of the responding States (48 of 61) supported ICAO taking this action.

The Conclusions and Recommendations of ATConf/6 were that the growth and expansion of air cargo services is beneficial for the sustainable development of air transport, and contributes to global trade and economic development; the distinct features of air cargo services, especially all cargo operations, need to be given due consideration by States when making air service arrangements; and ICAO guidance on liberalization of air cargo services remains relevant, and its use by States should be encouraged. ICAO can play an important role in facilitating further liberalization in this respect, including the development of a multilateral approach. Accordingly ICAO recommended that

(a) States should give due regard to the distinct features of air cargo services when exchanging market access rights in the framework of air service agreements,

particularly in the context of all cargo operations, and grant appropriate rights
and operational flexibility so as to cater to the needs of these services; and
(b) ICAO should initiate the development of a specific international agreement to
    facilitate further liberalization of all cargo air service, as discussed.[80]

6.3.1   Economic Aspects of Security

At ATConf/6 ICAO brought to bear the inextricable link between economic
liberalization and the need to ensure security. The Organization advised the Conference that the current ICAO policies on security charges for airports were updated
based on the outcome of the *High-Level Ministerial Conference on Aviation
Security*, held in Montréal in 2002. The policies list eight aviation security
functions, including required equipment, facilities and personnel, as well as security control of passengers to training of security personnel (Doc 9082, Appendix
1 refers), all of which should be taken into account in determining the costs basis for
charging purposes of an airport operator. States are responsible for ensuring the
implementation of adequate security measures at airports pursuant to the provisions
of Annex 17—*Security* to the *Convention on International Civil Aviation*. States
may delegate the task of providing individual security functions to such agencies as
airport entities, aircraft operators and local police.

In accordance with ICAO policies on charges contained in Doc 9082 (paragraph
1 of the Foreword refers), the four key charging principles of non-discrimination,
cost-relatedness, transparency and consultation with users should be observed when
security charges are designed and implemented. In accordance with a recommendation adopted by the Conference on the Economics of Airports and Air Navigation
Services (CEANS 2008) and endorsed by the ICAO Council, States are encouraged
to incorporate such key charging principles in national legislation, regulation or
policies, as well as in air services agreements, in order to ensure compliance by
airport operators and air navigation services providers.

It is a State responsibility to determine the circumstances and the extent to which
the costs incurred in providing security facilities and services should be borne by
the State, the airport entities or other responsible agencies. With reference to the
recovery of security costs from the users, the following general principles should be
applied[81]:

(a) Consultations should take place before any security costs are assumed by
    airports, aircraft operators or other entities;
(b) The entities concerned may recover the costs of security measures at airports
    from the users in a fair and equitable manner, subject to consultation;

---

[80] ATConf/6-WP/14.

[81] Doc 9082 Section II Paragraph 7.

(c) Any charges for, or transfers of, security costs to providers, aircraft operators and/or end-users should be directly related to the costs of providing the security services concerned and should be designed to recover no more than the relevant costs involved;

(d) Civil aviation should not be charged for any costs that would be incurred for more general security functions performed by States such as general policing, intelligence gathering and national security;

(e) No discrimination should be exercised between the various categories of users when charging for the level of security provided. Additional costs incurred for extra levels of security provided regularly on request to certain users may also be charged to these users;

(f) When the costs of security at airports are recovered through charges, the method used should be discretionary, but such charges should be based on either the number of passengers or aircraft weight, or a combination of both factors. Security costs allocable to airport tenants may be recovered through rentals or other charges; and

(g) Security charges may be levied either as additions to other existing charges or in the form of separate charges but should be subject to separate identification of costs and appropriate explanation.[82]

The *High-Level Conference on Aviation Security* (HLCAS), held in Montréal from 12 to 14 September 2012 highlighted the importance of defining security measures that are effective, efficient, operationally viable, economically sustainable, and which take into account the impact on passengers. The HLCAS specifically discussed the sustainability of aviation security measures, and recommended that ICAO and its Member States take into account an outcome-based approach, and the need for improved passenger satisfaction, in the long-term development of Annex 17.

6.3.2 Cargo Security

The air cargo industry is a $60 billion business[83] and passenger-accompanied air cargo is a major profit maker accounting for approximately 15 % of the industry's overall revenue.[84] At a high level aviation security conference held by the International Civil Aviation Organization in September 2012, ICAO advised that civil aviation was estimated to grow by 6 % in 2013; 6.4 % in 2014; and 4.9 % on an annual basis until 2020, such growth being encouraging to the air transport industry from an economic perspective. At the same time, ICAO cautioned that this growth

---

[82] ATConf/6-WP/18.

[83] Hoffer (2008) at p. 146.

[84] *Id.* 166.

may not necessarily halt the persisting threat of unlawful interference with civil aviation and that the threat will evolve into a complex web of activity as new innovative methods of attack are conceived. This brought to bear the compelling need to address the growing vulnerabilities of the air transport system and the sustainability of aviation security.

In this context the conference noted that air cargo security, its sustainability and the need for innovative approaches to risk management were considered paramount.

The High Level Conference on Aviation Security was attended by over 700 participants representing 132 of ICAO's 191 member States, and 23 Intergovernmental and industry international and regional organizations.[85] The Conference focused on three overarching themes on cargo security, sustainability of security measures and innovation in addressing threats to aviation security.

The Conference recalled the events of 29 October 2010 when terrorists exploited vulnerabilities in the air cargo security system to introduce improvised explosive devices intended to destroy aircraft in flight, and endorsed certain principles on air cargo and mail security. One principle was that a strong, sustainable and resilient air cargo security system is essential and that the threat is to the air cargo system as a whole and risk-based consideration must be given to strengthening security measures across all aspects of the system, including enhancing the ability to recover from a major disruption. Another was that appropriate security controls should be implemented at the point of origin. Cargo and mail should come from a secure supply chain or be screened and, in either case, protected throughout the entire journey, including at transfer and transit points. At points of transfer, States should satisfy themselves that security controls previously applied to cargo and mail meet ICAO Standards. In doing so, they should avoid unnecessary duplication of security controls.

An important milestone in this regard was the recognition that air cargo advanced information for security risk assessment is a developing area that enhances air cargo security, particularly in the context of express delivery carriers such as FEDEX, UPS, DHL Express and TNT Express who carry around 30 million shipments daily, which typically contain high-value added, time-sensitive cargo. These carriers guarantee the timely delivery of these vast volumes of shipments, ranging from same-day delivery to 72 h after pick-up, virtually anywhere in the world. They operate in 220 countries and territories.

The conference noted that a real risk in the area of cargo and mail security would arise when an express delivery carrier experiences a technical problem in an aircraft and is forced to transfer cargo to a passenger carrier, in which instance strict supply

---

[85] The Conference was the first high-level ICAO meeting on aviation security since early 2002, when Member States adopted a new and highly successful global security strategy following the 9/11 attacks. It also supported the implementation of the Declaration on Aviation Security adopted at ICAO's 37th Assembly in 2010, a discussion of which will follow, and was the culmination of a series of regional consensus-building security events organized by ICAO during 2011 and 2012.

chain standards should be adhered so that the risk in the transfer of cargo could be obviated.

Participants agreed that it was essential that solid standards and mutual recognition programmes be in place in order to make sure that States all along an air cargo supply chain satisfy themselves that air cargo is secure, and so let it flow unimpeded. Such standards and recommended practices should allow for the speedy transit and transhipment of legitimate air cargo worldwide, through any combination of air routes and transit or transhipment points.

Another risk factor considered by the Conference was the insider threat and the need to implement 100 % screening of persons other than passengers. This includes personnel at the airport, visitors and others who do not carry a boarding card. The Conference was reminded that States needed to acknowledge that the roles of people other than passengers that are working in civil aviation can present particular vulnerabilities that should be addressed.

It was also noted that there is already an ICAO Standard which requires each State to ensure that persons other than passengers, together with items carried, being granted access to security restricted areas are screened; however, if the principle of 100 % screening cannot be accomplished, other security controls, including but not limited to proportional screening, randomness and unpredictability, should be applied in accordance with a risk assessment carried out by the relevant national authorities.

One of the recommendations put forward at the Conference was to establish more airport authorities with increased aviation security expertise. In this regard it was noted that comprehensive background checks of all personnel selected for hire/ employment at an airport need to be carried out by the relevant State's security agencies based on risk assessment. In addition, re-vetting of airport workers such as cleaners, duty free shop personnel, catering staff and concessionaires must be carried out frequently in order to mitigate collusion to commit acts of unlawful interference.

The author is of the view that if aviation security concerning cargo and mail were to be addressed in a results-based manner, there are two key areas that need to be enforced: sustainability and innovation. 'Sustainable aviation security' can be defined as the detection and prevention of, and response to and recovery from, acts or attempted acts of unlawful interference with civil aviation, utilizing means that can be sustained by the entity or entities responsible for the period of time required. It is worth noting a number of important inter-related policy principles and practices that can contribute to the achievement of sustainable aviation security. These and other means can, more broadly, support the development of a sound and economically-viable civil aviation system. The starting point for consideration of any security measure must be a risk assessment. Such risk assessments, carried out objectively by appropriate security authorities on a continuous basis and informed by available and relevant information, including security intelligence, help assure that new or revised security measures are justified, aligned with actual needs and are proportionate to the level of risk.

The Conference recognized that the sustainability of aviation security measures and arrangements is an important strategic issue for all entities with aviation security-related responsibilities. It noted that risk-based security measures, outcomes-focused security measures, rationalization of security measures, optimization of technology, mutual recognition of equivalence and one-stop security, harmonization, and preparedness for crisis events are policy principles and practices whose implementation can contribute significantly to the sustainability of aviation security measures and arrangements.

From the time aviation was used as a weapon of mass destruction on 11 September 2001, there have been 75 terrorist attacks on aircraft and airports worldwide which have resulted in 157 deaths up to the end of 2011. When one compares this statistic to other modes of transport, such as trains and buses, one notes that there have been approximately 2,000 attacks and about 4,000 deaths resulting. On this basis, aviation has been fortunate. However, one cannot be complacent. The terrorist anchors himself on the *Displacement Theory*,[86] moving from one mode of attack to another when the going gets bad. The 9/11 attacks on buildings turned to attacks on airports and then onwards to cargo. Examples go back to the eighties where, in the early eighties, aircraft were attacked through the cargo hold (recalling just three instances when aircraft of Air India[87] and PANAM[88] were blown up in midair as well as the attack on the Airlanka aircraft[89]) to using aircraft as weapons of mass destruction in 2001, to attacking airports in the nineties. Recalling the events of 29 October 2011 discussed above, one could argue that unlawful interference with civil aviation has turned full 360 degrees and has seemingly returned to attacks on cargo.

---

[86] The Displacement Theory argues that removing the opportunity for crime or seeking to prevent a crime by changing the situation in which it occurs does not actually prevent crime but merely moves it around. In the present context it is called target displacement where crime can be directed away from one target to another.

[87] On 23 June 1985, an Air India aircraft operating flight 183 on the Toronto–London–Delhi route was blown up by a bomb at an altitude of 31,000 feet (9,400 m), and crashed into the Atlantic Ocean while in Irish airspace.
A total of 329 people were killed, including 280 Canadians, 27 British citizens and 22 Indians.

[88] A PANAM aircraft operating flight 103 from London to New York on Wednesday, 21 December 1988, was totally destroyed by an explosive device killing all 243 passengers and 16 crew members. Large sections of the plane crashed into Lockerbie, in southern Scotland, killing a further 11 people on the ground.

[89] Airlanka's L 1011 Tristar was destroyed on the tarmac at Colombo Airport. On 3rd May 1986, Flight 512 had arrived at Bandaranaike International Airport from London Gatwick and was about to fly on to Male when an explosion ripped the plane in two. The flight carried mainly French, West German, British and Japanese tourists. 21 people were killed on the plane which included 13 - foreigners—of whom two were British, 2 West German, 3 French, 2 Japanese, one Maldivian and one Pakistani, 41 were injured.

6.3.3   The Risk Based Approach

The Conference viewed with approval the ICAO Risk Content Statement and endorsed the Declaration adopted at the 37th ICAO Assembly Session convened by ICAO from 28 September to 8 October 2010, which, *inter alia* called for the strengthening of security screening procedures, enhancement of human factors and utilization of modern technologies to detect prohibited articles and support of research and development of technology for the detection of explosives, weapons and prohibited articles in order to prevent acts of unlawful interference.[90] The Declaration also calls on all Member States to share best practices and information on a range of key aviation security matters, including threat-based risk assessments.

6.3.3.1   The Risk Content Statement

The Risk Content Statement submitted to the Conference by ICAO was entirely based on a risk based approach which advocated a robust methodology for national risk assessment. It aims at providing a description of the global risk picture; assisting States in their efforts to protect air transportation and prevent its use for unlawful acts; presenting high-level statements for an improved approach in creating and maintaining State national civil aviation security programmes; and assisting ICAO in improving Standards and Recommended Practices (SARPs) and guidance material. The Statement focused the attention of ICAO member States to Standard 3.1.3 of Annex 17 to the Convention on International Civil Aviation (Chicago Convention) which requires each State to keep under constant review the level of threat to civil aviation within its territory, and establish and implement policies and procedures to adjust relevant elements of its national civil aviation security programme, based on a security risk assessment carried out by relevant national authorities. The Conference, based on this fundamental premise, recognized that a reasonably designed risk-based approach is one by which States identify the criteria to measure potential criminal activities, principally from terrorism. The identification of risks permits States to determine and implement proportionate measures and controls to mitigate against each risk type.

The Statement exhorted member States of ICAO to share information based on the premise that in conducting a risk assessment, it is necessary to assemble information about the threat. Such information may come from a variety of sources, such as those relating to: actual incidents, including successful or unsuccessful attacks on aviation, which provide information on proven terrorist methodologies;

---

[90] The Declaration also calls for the development of enhanced security measures to protect airport facilities and improve in-flight security, with appropriate enhancements in technology and training; and development and implementation of strengthened and harmonized measures and best practices for air cargo security, taking into account the need to protect the entire air cargo supply chain.

closed sources, from primarily counter-terrorist intelligence, which may be gathered by intelligence, law enforcement and other agencies of States; and open sources, which may include publicly available information on unusual or suspicious occurrences and the availability of items that could be used for terrorist purposes, and any other information that may contribute to the threat picture.

The Statement also highlights the insider threat as being of compelling significance. It emphasizes the danger of vulnerability associated with insiders which may be considered greater if they have access to the last layer of security in a way that a passenger does not, and points out that the likelihood associated with insiders might be less if they have already been subject to vetting and selection procedures and/or screening. It cautions that the consequence of a threat associated with insiders might be greater if an insider has access deeper within the system. For instance, an insider could perpetrate a more credible and thus more disruptive hoax. In summary, the Statement explains that the methodology involves considering each role within the system and whether it offers a particular tactical advantage in relation to each threat type or whether it poses the same issues as passengers, and that, in applying this methodology, it is possible to consider insider vulnerabilities as part of an integrated risk assessment.

Finally, the Risk Content Statement identifies risk assessment as a process which evaluates risk by *threat identification*, i.e. identifying the threat scenario, consisting of a defined target (e.g. airport terminal or aircraft), as well as the means and method of possible attack (e.g. attack by passenger using an improvised explosive device, or attack by an insider using weapon, etc.); *likelihood* i.e. considering the probability of the threat occurring[91]; consequence—assessing the nature and scale of likely impacts associated with a successful attack, including consideration of human, economic, political and reputational factors (based on a reasonable worst-case scenario); *vulnerabilities*, i.e. evaluating the effectiveness and vulnerabilities of current security measures (i.e. security strengths and weaknesses of SARPs) in mitigating the potential threat scenario identified; and *residual risk* i.e. assessing the remaining risk of that type of attack being successfully carried out against that target, to enable a judgement to be made as to whether that is acceptable in risk management terms.

6.3.3.2 Capacity Building

The High Level Security Conference noted that capacity building on an international scale was critical to a risk based approach for air cargo security. In this regard it was recognized that an international capacity building strategy for air cargo and

---

[91] Blaise Pascal, in his book *Ars Cogitandi* states that fear of harm ought to be proportional not merely to the gravity of the harm but also to the probability of the event. It is also a fact of risk management that, under similar conditions, the occurrence (or non-occurrence) of an event in the future will follow the same pattern as was observed in the past. For a discussion on risk assessment and risk management see Ferguson (2008) at p. 188 and Bobbitt (2008) at pp. 98–179.

mail security would draw on the *ICAO Assistance and Capacity Building Strategy for Aviation Security*,[92] and allow for targeted assistance for States in need. This Strategy would be guided by the ICAO security audit results, where air cargo and mail security has been identified as a priority need. The Strategy would include a proposal to coordinate bilateral and multilateral capacity building initiatives region-ally, as well as amongst international organizations such as the World Customs Organization (WCO) and the Universal Postal Union (UPU) in order to align such initiatives, maximize limited resources, and avoid duplication of efforts. The development of such a strategy would also be in line with the *ICAO Comprehensive Aviation Security Strategy (ICASS) 2011–2016*[93] that was endorsed at the 37th ICAO Assembly in 2010.

In this regard the Conference viewed favourably the idea that any new arrangements must recognize that many donor states engage in aviation security capacity building for specific national interest reasons, generally related to the nature of flights into donor states. This is understandable and, in fact, is a concept which drives many bilateral aviation security efforts across the globe. In encourag-ing this capacity building to continue, the proposed framework seeks to better coordinate and inform its development by building on existing and future Govern-ment-to-Government arrangements with targeted industry-to-industry capacity building efforts, and using ICAO-sponsored capacity building where regional "gaps" in bilateral, multilateral and industry capacity building efforts are identified.

The reality that capacity building requires a long-term commitment and should be focused on "regular/repeated engagement", rather than the provision of one-off courses on an irregular basis, was recognized, together with the fact that effective capacity building takes years and will only succeed when issues of trust, mutual respect and culture are addressed and fostered on an on-going basis. The Confer-ence endorsed the development of an International Capacity Building Strategy specific to air cargo and mail security, to aid those ICAO member States that require assistance to adequately implement enhanced ICAO air cargo and mail security standards. This Strategy was to be aligned with the *ICAO Assistance and Capacity Building Strategy for Aviation Security*, and avoid duplication of efforts. It also encouraged all ICAO Contracting States to further support the Secretariat in its efforts to provide capacity building assistance based on USAP audit results,

---

[92] The *ICAO Assistance and Capacity Building Strategy for Aviation Security* recognizes that there is a strong need as well as demand on ICAO for capacity building initiatives and that such assistance for States is vital for success. The Strategy further observes that resource demands on the organization require ICAO to manage priorities and expectations.

[93] The ICASS comprises seven strategic focus areas, one of which is titled Promoting Global Compliance and Establishing Sustainable Aviation Security Oversight Capability of States. The other six strategic focus areas are: addressing new and existing threats; promoting innovative, effective and efficient security approaches; promoting the sharing of information amongst and within member States to raise awareness of threats and aviation security trends; improving human factors and security culture; promoting the development of mutual recognition for aviation security processes; and emphasizing the importance of security amongst States and stakeholders.

subject to the consent of the State(s) receiving assistance, focusing on air cargo and mail where it has been identified as a priority; and urged other entities within the air cargo environment to continue taking action to effectively secure those sections of the supply chain in which they operate.

6.3.3.3   Insider Threats

A third element addressed within the parameters of the risk based approach was the insider threat and the need for screening of persons other than passengers. The Conference considered as the basis for discussion Standard 4.2.6 to Annex 17 to the Chicago Convention which, through Amendment 12 to the Annex (which became applicable on 1 July 2011), states that each Contracting State is required to ensure that persons other than passengers, together with items carried, being granted access to security restricted areas are screened; however, if the principle of 100 % screening cannot be accomplished, other security controls, including but not limited to proportional screening, randomness and unpredictability, shall be applied in accordance with a risk assessment carried out by the relevant national authorities.

The above provision notwithstanding, the Conference recognized that it was indeed very difficult to preclude, detect and face an act of unlawful interference carried out with the internal support of persons who have access to security-restricted areas, even though such persons may have had their records verified. The danger and risk were compounded by the fact that such persons usually have access to sterile lounges and other security-restricted areas where they have the opportunity to mingle with passengers and therefore could well interfere with their carry-on baggage and/or the checked baggage already inspected. They also have access to aircraft during ground and pre-flight services. One participating State suggested that in all access control points, conditions should be created securing that 100 % of persons who are not passengers, as well as the articles transported, are subject to security inspections with whatever of the different means available for this purpose, including manual inspection.

Some national practices submitted to the Conference were: (a) supervise or accompany daily/seasonal workers in the restricted security area; (b) closely examine all officials, employees or staff entering the restricted security area; closely examine all Janitors before permitting them to enter the aircraft; oversee the restricted security areas and facilities related to flight operations by patrolling periodically or continuous surveillance using CCTV; inspect all cabin carry-on, baggage and cargo as well as food (catering items) and equipment required and sold in flight (stores) and watch them before and during the loading onto the aircraft; oversee the process of boarding passengers and loading of goods; aircraft security check before departure (pre-flight security check); supervise, control and update the permit issuance and use of appropriate entry of airport, including applying background checks and stop list procedure; implement security awareness training for all airport pass applicants; carry out internal and external quality control regularly based on risk assessment; and be aware of religious, social and cultural approach among stakeholders.

The Conference recognized that the sustainability of aviation security measures and arrangements is an important strategic issue for all entities with aviation security-related responsibilities and that risk-based security measures, outcomes-focused security measures, rationalization of security measures, optimization of technology, mutual recognition of equivalence and one-stop security, harmonization, and preparedness for crisis events are policy principles and practices whose implementation can contribute significantly to the sustainability of aviation security measures and arrangements.

The need for each State to carry out continuous risk assessments as a preliminary measure was considered paramount for the sustainability of security. One view was that aviation security has to be sustained in a balanced manner so that, on the one hand, applying security measures to mitigate identified threats, and on the other hand, the essential task of facilitating operations, passengers' experience and trade could be ensured. Security should not accumulate layer upon layer of controls and associated costs, but should rather ensure the sustainability of the system from the perspectives of cost, efficiency, and acceptability by passengers and air transport operators, which should be a central consideration when designing security processes. Another means of achieving sustainability is at transfer points where security controls are known to have been performed effectively at the point of origin. The Conference took note of the fact that, in such instances, the concept of 'One Stop Security' should be advanced, where ICAO Member States, by virtue of recognising the equivalence of each other's aviation security regimes, can allow incoming passengers, baggage and cargo to transfer onto a connecting flight without being subjected, once again, to the same security controls as at the point of origin. The conclusion of such 'One Stop Security' arrangements remains an issue to be addressed Member State to Member State.

One proactive suggestion towards achieving and retaining sustainability was to follow the practice of reciprocal acceptance of equivalent security measures across the board, with due regard to the principle of host State responsibility, as envisaged by the Chicago Convention. In that respect, it was suggested that the need for any one State to require extra security measures of another State can be avoided by working together to align international requirements to the global threat environment. A further recommendation was that this approach should be reflected in Annex 17 to the Chicago Convention which deals with the subject of security.

The Conference was called upon to endorse a coordinated response to security incidents and threats whereby States could collectively accept, without derogating a State's freedom to take its own measures, the measures adopted by one State as a global norm if that norm were to be consistent with the Standards and Recommended Practices of Annex 17 and provided such recognition was accepted as such by ICAO.

Incontrovertibly, innovation in air cargo security lies in two areas: advancement of technology; and intelligence. In the field of air cargo security the ICAO

Conference showed a marked deficiency of discussion.[94] It is submitted that, critical to a discussion of technology and innovation is the subject of supply chain security. Preeminent among technological progress is the need to establish basic security packaging mandates for shippers. Cargo is either being flown or stored at any given point in time and therefore both phases must be covered in the tracking and identification of cargo. Hoffer recommends:

> Courier boxes and envelopes supplied by carriers should be required to have an original number and (if possible) a tamper-evident seal and markings (tied to the bill of lading), so that it is harder to replace a package with a similar box. Recipients would have the ultimate responsibility to compare manifest numbers with packages before accepting them.[95]

Air cargo can be loaded individually or in bulk form from one kilogram to a weight of several tons and can be loaded on various platforms such as unit load devices (ULDs) crates and assembled pallets. Several technologies are used at present in ensuring cargo security. These may vary between explosive detection devices, explosive trace detection computer aided tomography and X-ray, in addition to certified canine teams. The Transportation Security Administration (TSA) has identified such advanced technologies as XR/PFNA X-ray systems with pulsed fast neuron analysis; pressure activated sampling systems, quadruple resonance and miniature explosive and toxic chemical detector utilizer sensors.

In the context of military intelligence, the author submits that as a mirror reflection of the "known shipper" and "known consignor"[96] practice, military intelligence be employed to track and identify unknown consignors as well as insiders. Taking into consideration the aircraft bombings that have taken place (some of which have been discussed in this article) it is fair to conclude that most of these attacks were perpetrated by groups of incendiary persons. Military intelligence, which essentially is information relating to the armed forces of a foreign country that is significant to the planning and conduct of another country's military doctrine, policy, and operations, largely penetrates such groups and could be effectively used to take pre-emptive and preventive measures against threats to air cargo security.

---

[94] The conference discussed at some length technology pertaining to passenger screening and identification. It encouraged States to exchange data on the Public Name Record (PNR) and Advance Passenger Information (API) and to enhance the use of Machine Readable Travel Documents (MRTDs) and join the Public Key Directory (PKD). For a discussion on these issues see Abeyratne (2005b), pp. 170–174. See also Abeyratne (2005a), pp. 255–268. Also, see generally, Abeyratne (2010), pp. 121–156.

[95] Hoffer, *supra*, note 83 at 162.

[96] The concept of known shipper and known consignor is gaining fast popularity as a regular method of ensuing cargo security. Annex 17 to the Chicago Convention defines a known consignor as a consignor who originates cargo or mail for its own account and whose procedures meet common security rules and standards sufficient to allow the carriage of cargo or mail on any aircraft. These categories of persons are trusted shippers who have gained a good reputation with the carrier through a history of transactions and business dealings. By using this system, unknown shippers and consignors can be sent through more cautions and rigorous screening processes. For more details see Buzdugan and Flouris (2010) at p. 174.

It is eminently clear that the glue that binds the elements discussed above, including those that relate to the global supply chain, is law and practice. These are already in place in principle. For instance, Standard 4.6.1 of Annex 17 requires each Contracting State to ensure that appropriate security controls, including screening where applicable, are applied to cargo and mail, prior to their being loaded onto an aircraft engaged in passenger commercial air transport operations. The operative words here are "security controls" which brings to bear the reality that different States could have different security controls[97] and that they should be harmonized in ensuring supply chain security and global security standards. Screening and examination of cargo and mail are paramount to this consideration.

Standard 4.6.2 requires that each contracting State establish a supply chain security process, which includes the approval of regulated agents and/or known consignors, if such entities are involved in implementing screening or other security controls of cargo and mail. A regulated agent is defined in the Annex as an agent, freight forwarder or any other entity who conducts business with an operator and provides security controls that are accepted or required by the appropriate authority in respect of cargo or mail. There are five other provisions under Chapter 4.6 of Annex 17 pertaining to: the protection of cargo and mail from unauthorized interference from the point of screening or other security controls being applied until the departure of the aircraft[98]; the non-acceptance of cargo or mail by operators unless it is confirmed that screening or other procedures have been applied and conformed by a regular agent[99]; the appropriate screening of catering, stores and supplies intended for carriage by air[100]; the appropriate screening of merchandise and supplies introduced into security restricted areas[101]; and the fact that the security controls mentioned above have been implemented on the basis of a security risk assessment carried out by relevant national authorities.[102]

If laws and practices are the glue that keeps air cargo security together, political will is the fuel which will ignites its progress and development. The thrust of political will essentially lies in a security culture that must be visible in every State. A security culture would make States aware of their rights and duties, and, more importantly, enable States to assert them. Those who belong to a security culture also know which conduct would compromise security and they are quick to educate and caution those who, out of ignorance, forgetfulness, or personal weakness, partake in insecure conduct. This security consciousness becomes a "culture" when all the 191 member States of ICAO as a whole make security violations

---

[97] Annex 17 defines a security control as "a means by which the introduction of weapons, explosives or other dangerous devices, articles or substances which may be used to commit acts of unlawful interference can be prevented.

[98] Standard 4.6.3.

[99] Standard 4.6.4.

[100] Standard 4.6.5.

[101] Standard 4.6.6.

[102] Standard 4.6.7.

socially and morally unacceptable within the group. In building a security culture within ICAO member States it is imperative that consideration should also be given to the development of a process for ensuring that all Member States are notified when deficiencies identified during the course of an ICAO security audit conducted under the Universal Security Audit Programme (USAP) remain unaddressed for a sustained period of time. A notification process could involve the use of information which does not divulge specific vulnerabilities but enables States to initiate consultations with the State of interest to ensure the continued protection of aviation assets on a bilateral basis. States have to adopt a security culture that admits of an overall approach to the threat as a potential harm to humanity. This should inevitably include strict adherence by States to the provisions of Annex 17.

6.4 Economic Regulation of Air Navigation Services Providers

At ATConf/6 the Civil Air Navigation Services Organization (CANSO) played a significant role in bringing to the forefront the important link between air transport and the provision of air navigation services. CANSO stated that establishing and maintaining effective and mutually beneficial relationships with customers was simply good business. Accordingly, CANSO and other key stakeholders recognised that in order to meet the challenges of improving aviation safety, service, cost-effectiveness, and environmental/fuel efficiency, it is vital that they work with their customers and other stakeholders in the industry.[103] The key emphasis was customer relations and CANSO Members believed that effective customer relations and consultation should be founded on the following fundamental principles, which can be applied to any kind of customer, for example any category of airspace user or airport operator put forward the following elements for consideration:

(a) **Mutual trust & understanding**: Relationships should be founded in an environment of mutual trust and understanding. It is important that ANSPs openly share information related to their future plans and current performance with customers on a regular basis. It is equally important for customers to share information about their plans so that ANSPs can take this into account when developing their own future plans;

(b) **Early involvement of customers**: Customers who may be affected by a policy or plan should be engaged from an early stage, and throughout its development. Sufficient time should be allotted for ANSPs to consult with their customers;

(c) **True consultation**: Effective consultation should ensure that customers have the opportunity to have their say and know that their interests have been considered in policy decisions;

---

[103] ATConf/6-WP/71.

(d) **Clear & focused agenda**: Issues should be discussed informally with customers to establish an agenda and focus the formal consultation process;

(e) **Clarity of proposals, the rationale and their impact**: Customer consultation should make clear the nature of proposals; the parties most likely to be affected; the business case for proposals; specific questions on which feedback is requested; and the time schedule for responses;

(f) **Clear communication**: Consultation documents should be concise, clearly laid out and should make use of simple language wherever possible;

(g) **Tailored processes**: Consultations should use methods appropriate for the subject matter including seminars, working groups and oral briefings as well as standard written consultation exercises;

(h) **Constructive participation**: Reasoned responses should be provided by interested parties; responses should be acknowledged and all respondents should automatically receive copies of the final decision document; and

(i) **Convincing rationale shared following final decisions**: Decision documents should contain clear reasons for the chosen outcome(s), and should show how responses have been taken into account.

CANSO stated that many factors contributed to the success of the link between the consumer and the service provider which included: level of development of the parties involved in the relationship; the complexity of the industry environment; cultural influences; and characteristics of the customer and ANSP. CANSO summed up by saying that in order to achieve superior air navigation services performance, the focus of policy-making needs to be on management behaviour and how it is best influenced. Superior performance can best be achieved by a governance structure that clearly articulates the objectives for which the ANSP is to be governed and holds management accountable for results.[104]

Service priority policy was an important discussion at ATConf/6 where ACI, IATA and CANSO were in unison in their presentation that service priority policy needs to take into account several important issues or principles for it to succeed in its objective. Firstly, while the goal of any enabling policy must be to improve overall system capacity and efficiency, there may be instances when granting priority to equipped capable flight does not produce an improvement in system performance. It may only be after a certain percentage of aircraft have the capability needed that overall benefit is achieved. However, it should be recognised that in order to induce sufficient number of aircraft to equip with the new capability it will be necessary to provide benefit to the already capable flights before this threshold is met. When operational benefits to more capable flights are not substantial in a first phase, economic/financial incentives—also within the charging system—may be even more important to encourage a fast introduction of new equipment to reach this threshold.

---

[104] ATConf/6-WP/73.

Second, the three Organizations argued that it may be the case that the only practical way to direct performance benefits to more capable flights aircraft is to induce a degradation in performance to some non-capable flights some of the time, as in imposing a hold on an non-capable flight. It must be accepted that those flights that are not capable will be relatively disadvantaged over the long-term. This being said, it may be part of the deployment strategy and business case to determine what magnitude and frequency differences in services are necessary to achieve the required system performance.

Third, they submitted that it must be considered that flight capability upgrade decisions, and the related training and certification of flight crews, need to be supported by a positive business case. Operators should expect a return on their investment according to a realistic and agreed timeframe, and the direct benefits received through financial and operational capability incentives should therefore be considered an integral part of the analysis.

Fourth, service priority policy must be targeted and tied to a specific service performance improvement objective. A basket of operational and/or financial measures exist or can be created, and choices will need to be made as to their suitability in meeting the performance improvement objective. The measure to be applied will depend on the fleet capability mix, aircraft equipage and crew capability, the type of operating environment (terminal versus en route or oceanic), and the air traffic planning phase (strategic, pre-tactical and tactical) that is being considered. If operational measures will have a positive financial/economic impact, such as less fuel burn and saved flight time, operational measures are more easily introduced in the en route or oceanic environment, while it may be needed to introduce financial/economic measures in congested or complex terminal operating environments as a first step before operational benefits become significant at a larger scale—until such time that operational measures can be more easily applied at a larger scale in that particular environment.

Fifth, performance metrics need to be defined that will measure and evaluate the implementation of the service priority policy. A performance baseline should be set in relation to the intended system performance goals.

Sixth, operational and/or financial measures need to be considered in the early planning phase, especially if multiple States/ANSPs are involved, and introduced in a transparent and non-discriminatory manner in collaboration with airspace users. Further, financial measures need to be offered on a temporary basis, until such time when more capable flights form a distinct majority and overall system performance improvements have been achieved.

Seventh, if direct financing of airborne equipment is provided by the State which is implementing new systems to the airspace users under their registration, the benefits versus effects it might have on fair competition for international aviation need to be assessed. Lastly, but most importantly, a collaborative planning process involving all relevant stakeholders (regulatory authorities, ANSPs, airlines and other aircraft operators, airports, military and security organisations, pilots and air traffic controllers) is the best path to success in the design and implementation of service priority policy.[105]

The Chicago Convention, in its vision and wisdom, incorporates various provisions regarding the provision of air navigation services by States to aircraft flying over their territories. Firstly, the Convention guarantees, through provisions included in Chapter XV, that States which are unable to provide air navigation services to aircraft will be assisted. Secondly, Article 15 of the Convention assures airlines that every airport in a Contracting State that is open to public use by its national aircraft shall also be open under uniform conditions to the aircraft of all the other Contracting States. The conditions are deemed to apply to the use, by aircraft, of every Contracting State of all air navigation facilities, including radio and meteorological services, which may be provided for public use for the safety and expedition of air navigation. Charges levied for such services are deemed by Article 15 to be anti-discriminatory whereby aircraft are not to be charged for airports and air navigation services provided to them at a rate higher than those levied on the national carrier of the State which provides the service. To this end, Article 28 of the Convention obligates Contracting States to provide, as far as practicable in their territories, airports, radio services, meteorological services and other air navigation facilities to facilitate international air navigation according to Standards established pursuant to the Convention.

The tightly-set legal parameters of the Chicago Convention, particularly the assurance of air navigation services on an equal and non-discriminatory basis, are relevant in the twenty-first century, where service providers and airline operators have to collaborate in ensuring a seamless global air navigation system. Modern technology offers sophisticated air-ground data communications by VHF (very high frequency) and satellite, assisted by precise navigation by inertial/GNSS and computing in air traffic services. These will be used in the negotiation of dynamic user preferred routes offering various alternatives to airline operators which provide fuel and time savings. However, such preferences for flight profiles and uses thereof will be subject to growing air traffic demands which have to be cautiously assessed. This imposes an added burden on both the service provider and airline operator. Judgment and interpretation will be critical factors in this process, an inevitable corollary of which will be the need to examine legal aspects of the modern seamless air traffic management system.

Responsibility of States for the provision of air navigation services in their territories is founded in principles contained in Article 28 of the Chicago Convention of 1944. It must be noted that this is not an absolute obligation as the State is called upon to provide such services only in so far as it finds practicable to do so. In order to cover an eventuality of a State not being able to provide adequate air navigation services, the Convention imposes an overall obligation on the Council of ICAO in Article 69 to the effect that the Council shall consult with a State which is not in a position to provide reasonably adequate air navigation services for the safe,

---

[105] ATConf/6-WP/74.

regular, efficient and economical operations of aircraft. Such consultations will be with a view to finding means by which the situation may be remedied. Article 70 of the Chicago Convention even allows for a State to conclude an arrangement with the Council regarding the financing of air navigation facilities and the Council is given the option in Article 71 of agreeing to provide, man, maintain and administer such services at the request of a State.

The provision of air navigation services are mainly regulated by three Annexes to the Chicago Convention, namely Annex 2 (Rules of the Air), Annex 3 (Meteorological Service for International Air Navigation) and Annex 11 (Air Traffic Services).[106] Of these, compliance with Annex 2 is mandatory[107] and does not give the States the flexibility provided in Article 38 of the Chicago Convention to register differences from any provisions of the Annex.

References

Abeyratne RIR (1991) Taxation in the field of international air transport – legal aspects. Air Space Law XVI(3):106–118

Abeyratne RIR (1992) Law making and decision making powers of the ICAO council – a critical analysis. Zeitschrift Luft Weltraumrecht 41(4):387–394

Abeyratne RIR (1993) Air transport tax and its consequences on tourism. Ann Tourism Res 20 (3):450–460

Abeyratne RIR (1995) Recent developments in taxation of air transport – the ICAO-IATA symbiosis. Air Space Law XX(2):48–59

---

[106] Article 54 (l) of the Chicago Convention stipulates as a mandatory function of the Council the act of adopting, in accordance with Chapter VI of the Convention, international standards and recommended practices (SARPs) and for convenience designate them as Annexes to the Convention. Article 37 of the Convention reflects the areas in which SARPs should be developed and Annexes formed. Article 38 obliges contracting States to notify ICAO of any differences between their own regulations and practices and those established by international standards or procedures. The notification of differences however, does not absolve States from their continuing obligation under Article 37 to collaborate in securing the highest practicable degree of uniformity in international regulations, standards, and procedures.

[107] In October 1945, the Rules of the Air and Air Traffic Control (RAC) Division at its first session made recommendations for Standards, Practices and Procedures for the Rules of the Air. These were reviewed by the then Air Navigation Committee and approved by the Council on 25 February 1946. They were published as *Recommendations for Standards, Practices and Procedures – Rules of the Air* in the first part of Doc 2010, published in February 1946. The RAC Division, at its second session in December 1946 to January 1947, reviewed Doc 2010 and proposed Standards and Recommended Practices for the Rules of the Air. These were adopted by the Council as Standards and Recommended Practices relating to Rules of the Air on 15 April 1948, pursuant to Article 37 of the Convention on International Civil Aviation (Chicago, 1944) and designated as Annex 2 to the Convention with the title *International Standards and Recommended Practices – Rules of the Air*. They became effective on 15 September 1948. On 27 November 1951, the Council adopted a complete new text of the Annex, which no longer contained Recommended Practices. The Standards of the amended Annex 2 (Amendment 1) became effective on 1 April 1952 and applicable on 1 September 1952.

Abeyratne RIR (1996) Legal and regulatory issues in international aviation. Transnational Publishers Inc., New York

Abeyratne RIR (1997) The settlement of commercial aviation disputes under the general agreement on trade in services and the ICAO council – a comparative analysis. International Trade Law and the GATT/WTO Dispute Settlement System. Kluwer Law International, London, pp 395–412

Abeyratne RIR (2001) Revenue and investment management of privatized airports and air navigation services: a regulatory perspective. J Air Transport Manag 7:217–230

Abeyratne RIR (2005a) The e-passport and the public key directory – consequences for ICAO. Air Space Law XXX(4–5):255–268

Abeyratne RIR (2005b) The use of information contained in airline passenger name record – some issues. Commun Law 10(5):170–174

Abeyratne RIR (2010) Aviation security law. Springer, Heidelberg, pp 121–156

Bobbitt P (2008) Terror and consent: the wars for the twenty first century. Knopf, New York

Boiteux M (1964) Peak-load pricing. In: Nelson JR (ed) Marginal cost pricing in practice. Prentice-Hall, Englewood Cliffs

Brownlie I (1990) Principles of public international law, 4th edn. Clarendon Pres, Oxford, p 691

Buzdugan M, Flouris T (2010) Regulation of air cargo security at the international level. In: Thomas AR (ed) Supply chain security. Praeger Security International, California

Ferguson N (2008) The ascent of money. The Penguin Press, New York

Hinshaw H (1939) The protection of aviation from inequitable taxation. J Air Law 9:75–94

Hoffer E (2008) Air cargo. In: Thomas AR (ed) Aviation security management, Chapter 9. Praeger Security International, California

MacKenzie D (2008) ICAO, a history of the International Civil Aviation Organization. University of Toronto Press, Toronto

Shaw MN (2003) International law, 5th edn. Cambridge University Press, Cambridge, 110

Tammes AJP (1958) Decisions of international organs as a source of international law. HR 94:265

Chapter 7
The Black Swan Effect

The metaphor of the black swan posits that, amidst the presumption that all swans are white, a black swan may emerge that did not exist and could not have existed prior to its actual appearance. In my book *Administering the Skies – Facing the Challenges of Market Economics*[1] I postulated that there is an integral link between the way in which the air transport product is influenced by market economics and the fluctuations of the industry's fortunes. Global economic trends, looked at in retrospect have affected the vicissitudes of air transport. Air transport, as any other business is affected by the business cycle, which has been commented upon as:

> The business cycle is a pulse common to almost all sectors of economic life and to all capitalistic countries. Movements in national income, unemployment, production prices and profits are not so regular and predictable as the orbits of the planets or the oscillations of a pendulum, and there is no marginal method of forecasting the turns of business activity.[2]

In 2009 there was unprecedented market volatility in the airline industry, after the boost enjoyed by the industry with the emergence of new carriers, and the subsequent fall as the global financial crisis took control. This trend resulted a few years later in market consolidation and leverage partnerships in the airline industry. Cost rationalization became a by word. This all goes to show the fragility of the industry. Nassim Nicholas Taleb, the bestselling author of *The Black Swan*, in his book *Antifragile* introduces the reader to the interesting and well-reasoned concept called "*Antifragile*". He states that any system which depends on predictability and presumption is fragile and that "some things benefit from shocks and they thrive and grow when exposed to volatility, randomness, disorder and stressors". According to Taleb black swans (which as we all know are a rarity) are large-scale unpredictable and irregular events which can either devastate those that are fragile and dependent on a certain rigid stability, or energize risk takers and flexible persons into action. Antifragility is associated with risk taking and anticipating the unthinkable.

---

[1] Aracne Editrice: Rome, 2012.

[2] Samuelson (1948) at p. 408.

At an interview with CNN in early 2013, Taleb gave the example of Egypt, which caved in under an uprising because they had depended on the only thing they knew—a dictatorship—and were so fragile as to disintegrate into chaos after the fall of the dictator. Another example cited is China, with an administration of appointed persons where there is no backup system if the administrative system it solely depends on, somehow fails. On the other hand he cites Switzerland as antifragile and survivalist, should political change happen there.

He calls those who comprise a fragile society *Fragilistas* whom he identifies as follows:

> The fragilista belongs to that category of persons who are in suit and tie, often on Fridays; he faces your jokes with icy solemnity, and tends to develop back problems early in life from sitting at a desk, riding airplanes, and studying newspapers. He is often involved in a strange ritual, something commonly called "a meeting". Now, in addition to these traits, he defaults to thinking that what he doesn't see is not there, or what he does not understand does not exist. At the core, he tends to mistake the unknown for the non-existent.[3]

One could arguably conclude on the aforementioned basis that we were fragilistas at 9/11, fragilistas when 20 little ones and 6 adults were massacred with an assault rifle by a deranged killer in an elementary school in Connecticut in December 2012, and fragilistas when a 23 year old woman was brutally raped and killed in India around the same time. We were in our suits attending meetings, snug as bugs in a rug until these events occurred. We were naive rationalists who believed in the stability of an assumed security.

And so it is with the airline industry; we will not be able to foresee or predict exactly the totally unpredictable. One commentator cites the aftermath of black Monday in 1987 where budget problems arose in the United States the economy of which went into a recession, in 1991 in Finland which went into recession when it lost its major business partner Russia, as did Australia and New Zealand. The fall back on the airline industry was evident. Major carriers such as Pan Am and Eastern Airlines disappeared, as did Air Europe across the pond, while Cathay Pacific, British Airways and Singapore Airlines stood steadfast, and in fact made healthy profits.[4] There is an inscrutable trend in the global economy that vacillates between prosperity and retrogresses once every 10 years or so, which has a ricochet effect on the airlines.

The foregoing discussion notwithstanding, it would not be totally justifiable to blame the bad fortunes of the airline industry solely on the ominous emergence of the black swan. Perhaps one could say that the airline industry blamed its misfortunes too much on the black swan. For instance, the industry has been quick to shift responsibility on external factors, such as contagious diseases or syndromes, acts of terrorism, rising fuel costs and the increasing burden of airport charges. The industry has over supplied capacity and as a consequence had to reduce fares to maintain load factors. The manufacturers love this, as well as the

---

[3] Taleb (2012) at p. 9.

[4] Clark (2010) at pp. 8–9.

airlines which order aircraft leveraging their mobility in the asset based financing markets. Government policy does not help as, in some instances as in the Middle East, airlines are pressurized to buy more and more aircraft in order to compete with each other. While this may be good for promoting connectivity, it bodes ill for the sustenance of a balance between supply and demand

7.1 Epiphenomena

Nassim Nicholas Taleb introduces the term "epiphenomena"[5] which is a causal illusion that attributes qualities to a concept, object or person which\who gives the illusion that it\he is responsible for a particular effect. An example cited is a compass on the bridge of a ship which would give the illusion that it is actually guiding or directing the ship, whereas in reality it is merely reflecting the direction of the ship. When one looks at ICAO as the compass and air transport as the ship, it is not difficult to understand this concept. One does not observe ICAO without observing the direction of air transport and one has the illusion that ICAO is actually causing air transport to develop. Instead, air transport has weathered the storm of fluctuating economic fortunes without the guidance of ICAO. In other words, the air transport industry has found ways to circumvent or confront the black swan by being antifragile. ICAO remains an epiphenomenon that gives the causal illusion that it has a "leadership" role in regulating air transport. This notwithstanding, one must hasten to add that ICAO has eminently directed aviation in technical issues such as the machine readable travel document, aviation system block upgrades.

ICAO's self-imposed inability to direct the regulation of air transport in the economics field is patent in its submission to ATConf/6 in March 2013 when it said on the subject of fair competition in international air transport:

> In fact, some air services agreements refer explicitly to the principle of a level playing field by noting that "where there is not a level competitive playing field for airlines, potential benefits deriving from competitive air services may not be realised". However, it must be recognized that there is currently no commonly accepted definition of the conditions constituting a "level playing field". It is unlikely that consensus on a comprehensive definition can be achieved at this time, given the widely different circumstances of States and their aviation sectors, including such fundamental issues as state ownership, policies on maintenance of national air carriers and airport development, and widely divergent State policies on taxation, labour regulation, bankruptcy, and health insurance.[6]

While this may be a true statement, it is unfortunate that ICAO stops at saying it is unlikely that consensus on a comprehensive definition can be achieved at this time, given the widely different circumstances of States, without offering some

---

[5] Taleb (2012) at p. 197.
[6] Fair Competition in International Air transport, ATConf/6-WP/4, at 3.

compelling direction as to what States should do about unifying a system of fair competition. This is a typical example of an epiphenomenon—a causal illusion that ICAO is actually showing the way to States to achieve fair competition whereas all that it is doing is to indulge in platitudes. ICAO admits that it should go beyond being illusory by recommending *inter alia* that ICAO should develop a set of core principles on fair competition in international air transport. Why does ICAO need the blessing of its member States to do this? Shouldn't this have been done ages ago—at least within the past two decades? And what will the status of these core principles be? How would they fit into the new Strategic Objective which requires ICAO to achieve "Economic Development of Air Transport: Foster the development of a sound and economically-viable civil aviation system"?

In another area relating to safeguard measures in liberalization ICAO refers to the *"Study on the Essential Services and Tourism Development Route Scheme"* (ESTD) conducted by ICAO and UNWTO and makes the disconcerting statement that:

> Although there is a lack of data on cases where the ESTD scheme has actually been applied by States at the international level, there is merit in continuing to provide ICAO guidance so as to raise State awareness and encourage use of the guidance. In this regard, ICAO should monitor developments and, as experience is gained, share information with States[7]

The question is, why conduct a study if it cannot be implemented through a decision taken at a conference such as ATConf/6? How is it that this cannot be accomplished through Appendix A Section IV of Assembly Resolution A37-20 (Consolidated statement of continuing ICAO policies in the air transport field) which requests the Council to: *"continue to exert a global leadership role in facilitating and coordinating the process of economic liberalization..."*[8]? What is the purpose of a leadership role if it only ascribes ICAO to develop policy and guidance? What exactly is meant by "foster the development of a sound and economically-viable civil aviation system"? How can leaders merely "foster"? Ironically, ICAO makes the statement that:

> the leadership role of ICAO in economic regulation of international air transport, and in the development of comprehensive policy guidance to assist States in the creation of a favourable regulatory environment for the sustainable development of air transport and the benefit of all stakeholders, is indisputable[9]

ICAO should exercise its leadership as a double headed sword bringing in the States as well into the picture. Certainly, ICAO has to lead in the process but not as a one sided organization. Chanakya, the advisor to King Chandraguptha in ancient India said "A good leader is one who makes his subjects fearless". This should in fact be ICAO's role, to lead States into competition. There are a few fundamental principles here. In the aftermath of the ATConf/5 of 2003 and the years to follow,

---

[7] Safeguard Measures for International Air Transport liberalization, ATConf/6-WP/3 at 3.
[8] Assembly Resolutions in Force (as of 8 October 2010) Doc. 9958, at III-7.
[9] Status of ICAO Policies in Air Transport Regulation, ATConf/6-WP/16 at 4.

States had to give cautious consideration to the application of the guidance provided both in the *Declaration* and the Conclusions and Recommendations of the Conference. The future for many airlines were fraught with overcapacity and falling yields. These negative factors were aggravated by high fuel costs. Liberalization has intensified competition, giving rise to even more dominant and powerful alliances and there is a real and tangible need for governments and regional blocs such as the European Community to enforce and regulate anti-competitive conduct of air carriers. Smaller and less competitive carriers, particularly those State owned, have had to stringently cut costs and depend on the regulator to keep the rise of fuel costs at a minimum.

Globalization of competition was one of the key messages of the Conference which brought with it the important message for governments to rethink their strategies with regard to the air transport strategy as an integral part of their national competitiveness. Any agreement to bring in an aspect of trade within a liberalized framework is generally a pro-active measure, which brings to bear the willingness and ability of the governments to face trading issues squarely in the eye. However, any agreement for trading benefits would be ineffective without the element of competition. The essential requisite for success in trading relations is competition, which in turn leads to national prosperity. A free trade agreement is merely the catalyst in the process.

The message ICAO should send its member States is that national prosperity is created, not inherited. Although national resources are a States' assets, the prosperity of a nation does not necessarily emerge solely from the natural endowments of the State concerned, nor from its labor resources, but rather from a certain localized process which engulfs economic structures, national values, culture and institutions. The essential catalyst to trade is national competitiveness.

7.2 The Low Fare Business Model

One way for the air transport industry to achieve antifragility or, in other words the defragilizing of the industry, is through the low fare—low cost business model which uses the point to point system as against the hub and spoke system used by most legacy carriers. The point to point system ensures the passengers of the lowest time of travel and ensures the airline lower costs in avoiding idle time for baggage handlers, gate staff and other ground crew who would wait for an aircraft with connecting passengers to arrive. The low fare model also yields the lowest airport charges. The low fare-low-cost model, which is the ultimate paradigm shift in the airline industry proved immensely successful in a short time. In 2009, they captured 41 % of the European market, 27 % of North America, 28 % of the Asia-Pacific market and 8 % of the South American market.[10]

---

[10] Dunn et al. (2009) at p. 56.

A trend was started in the beginning of the past decade where market conditions changed drastically where air transportation started growing twice as much as the general economy.[11] The result of this trend was a dramatic increase in the size of aggregate and individual aviation markets, resulting in turn in the emergence of the new breed of air carrier called the low cost carrier, who offers a simple and consequently low-cost service aimed at attracting customers with simple itineraries. The low cost carrier has grown to such significant lengths so as to compete with the largest established carriers in the world. The surge of liberalization and the healthy growth rate of 4–6 % of the aviation industry in the late 70s and thereafter for the next 20 years due to aggregate rises in gross domestic product, spurned an increased demand for travel, particularly by business travelers who had deeper pockets than did the average tourist. Major airlines contemporaneously launched a practice called "network management" using sophisticated computer technology and optimized business models. This was the period of matching expected demand and offered capacity through advanced quantitative analyses enabling major carriers to build global networks using the famous "hub and spoke" model. The trend encouraged network carriers to attract traffic to their designated hubs, even to the extent of creating multi-hub systems which were visible in the United States in the 1980s. However, intense competition between hub and spoke carriers created a standoff between major competitors who had to match others in optimizing desti-nation profiles even at the expense of productivity.[12] If carriers failed to match their competitors with their destination portfolios, they would be devoured by the competition through the computer reservation systems which penalized reductions in connectivity with reductions in bookings which in turn resulted in losses in revenue. This dilemma necessitated large network carriers to resort to other com-pensatory competition tools by forming partnerships and global alliances which were immensely popular in the mid 1990s and continue to flourish today in certain instances. These alliances would have been absolutely successful and thrived if existing regulatory restrictions on ownership and control air carriers were liberalized. However, since this did not happen globally, airline alliances failed to fully attain the desired cost reduction potential. The final nail in the coffin of the conventional legacy carrier was driven in late 2000 and throughout 2001 when the economic downturn which had started in an unprecedented manner grew worse and the fear factor spurned by the unfortunate events of 11 September 2001 saw a rapid decline in the demand for travel.

One distinct advantage of the low cost carrier service is its unwavering focus on the primary service, *i.e.* efficient and punctual carriage by air. The fact that the low cost carriers fly to secondary airports is proving attractive to passengers who increasingly prefer to go through airports which offer direct connections with less interactions than congested mega airport complexes. Low cost air travel also encourages the passenger—who would have chosen a cheaper alternate mode of

---

[11] Tretheway (2004), pp. 3–14 at p. 4.

[12] Franke (2004), p. 17.

transportation than the high priced fares of network carriers—to go by air instead of using other conventional modes of transport. Large full service network carriers are further confronted with the possibility, as demonstrated by Southwest which has operated low cost services for the past 30 years and has now established its own network of low cost destinations in the United States, that low cost services can drastically expand their own empires with their low cost based business models.

The point to point services offered by such low cost carriers as Ryanair, Easy Jet and Southwest are offered with operational efficiency and simplicity of service. These carriers use simple services and processes which in turn result in a simple, lean organization. This approach has been called "sustainable competitive advantage".[13] Usually, this advantage is gained by offering one class of service with open seating, no meals. This strategy has the dual advantage of cutting costs and simplifying cabin services. Aircraft types used are standardized (such as all 737 fleet in Southwest) which lowers maintenance and crew training costs to the minimum. Subscribing to one aircraft type also endears the operator to the manufacturer where the latter can give special deals to the former. Arguably, from a cost base perspective, the most strategic focus of the low cost carrier is on the use of secondary airports of small cities. These uncongested airports offer better facilitation of passengers and cargo by reducing circling time or in-air waiting to land or take off as well as ensure much lesser gate time than do congested airports of large cities. On a typical low cost flight, there is usually no interlining of baggage or passengers. All these factors result in the carrier's being able to lower its cost per available seat mile compared to its competition. Unfortunately, network carriers cannot attain this ultimate result because of the vicious cycle they are in—that they have to get rid of their networks to attain a sustainable competitive advantage of low cost but they are unable to, as their existence depends on their route networks.

The airline–airport relationship is an integral part of the low cost-legacy carrier equation. In addition to the cost factor, there is also the psychological factor affecting the passenger. Shorter walkways at airports and the minimum of confusion at gates and check-in points are critical advantages offered to the low cost carrier's passenger. Furthermore, secondary airports do not entail the usually difficult transfer connections, delays in the delivery of baggage, and even boredom.[14]

The most fundamental regulatory postulate which might apply to the low cost-legacy carrier equation is enshrined in the Chicago Convention which provides that one of the objectives of the International Civil Aviation Organization should be to prevent waste caused by undue competition.[15] To this end, at ATConf/5,

---

[13] Gillen (2004), p. 43.

[14] The Mintel International 2000 Study, conducted on views of passengers, found that 33 % of airport users found airports boring. The age group most affected was between 33 and 45. Of these, 29 % found London Heathrow to be boring, while 32 and 33 % had the same comment on Gatwick and Manchester respectively. See Barrett (2004) at p. 36.

[15] Article 44 (e).

Contracting States declared that liberalization of air transport must be accompanied by appropriate safeguard measures to ensure fair competition and effective and sustained participation of all States. This begs the question as to whether low cost carriers are indulging in unfair competition in terms of pricing. The ICAO Air Transport Conference agreed that States should give consideration to a model clause as an option for use at their discretion in air services agreements, which, *inter alia*, would consider that charging fares and rates on routes at levels which are, in the aggregate, insufficient to cover the costs of providing the air services to which they relate would be tantamount to an unfair competitive practice.[16] Of course this clause refers to charging air fares which are lower than the actual cost of the service provided. Therefore, if low cost carriers price their product lower than their cost base, they would be indulging in a practice inconsistent with this recommendation. The Conference also went on to identify as anti competitive, practices which involve the addition of excessive capacity or frequency of service, particularly if: by such practices as regular and sustained rather than sporadic or temporary; the practices in question have a serious negative economic effect on, or cause significant damage to, another airline; the practices in question reflect an apparent intent or have the probable effect, of crippling, excluding or driving another airline from the market; and behavior indicates an abuse of dominant position on the route.[17] The Conference recommended consultation between aeronautical authorities of State Parties if a conflict relating to this paragraph were to occur and subsequent resolution under relevant provisions pertaining to dispute resolution of ICAO. The Conference also considered important the fact that States need to carefully examine what elements of consumer interests in service quality have adequately been addressed by the current commercial practices of airlines (and service providers, if applicable) and what elements needed to be handled by the regulatory and/or Voluntary Commitment approaches. Some of the services recognized by the Conference under this heading were: availability of lower fares including fares on Web sites; reservation, ticketing and refund rules; check-in procedures; handling of compensation for flight delays, cancellation and denied boarding; baggage handling and liability; assistance regarding complaints; and assistance for disabled and special needs passengers.[18]

With regard to fares and rates, the ICAO Assembly, at its 32nd Session, held in Montreal on 25 September to 5 October 2001, adopted Resolution A32-17, which recognized that fares and rates to be established regarding international air transport must be fair and reasonable and designed to promote the satisfactory development of air services. The Resolution also recognized that States (or their governments) have a responsibility in fares and rates matters and requests the ICAO Council to

---

[16] See *Report of the Worldwide Air transport Conference*, Challenges and Opportunities of Liberalization, Montreal, 24-28 March 2003, ICAO Doc 9819, ATConf/5, 2003 at p. 37 [Para 2.3.3.2.1 (a)].

[17] *Id.* paras 2.3.3.2.1 (b) to (f).

[18] *Id.* para 2.4.3.1 (a).

keep under review the machinery for the establishment of international tariffs as well as the rules and conditions associated with international tariffs. Although this resolution is no longer in force, its thrust and spirit is embodied in Resolution A33-19 (Consolidated Statement of Continuing ICAO Policies in the Air Transport Field)[19] adopted at ICAO's 33rd Session of the Assembly (Montreal, 25 September-5 October 2001)which requests the Council to instruct the Secretary General to issue periodically a study on regional differences on the level of international air transport operating costs, analyzing how differences in operations and input prices may affect their levels and the impact that changes in costs may have on air transport tariffs.[20]

To this end, ICAO has published models of bilateral tariff clauses and identified determinative factors and mechanisms for developing tariffs[21] both for passenger and cargo carriage. ICAO has also identified the reasons for States to regulate international tariffs as being the need to ensure that their national carrier or carriers have a fair opportunity to operate international air services; to ensure that such carriers have a fair and equal opportunity to compete in providing international air services; to respond to the needs of uses of international air transport; and to promote competition in international air transport.[22] Low cost carriers emerged primarily because of the globalization of the air transport industry and market deregulation in most parts of Asia and Europe, which encouraged new enterprises to approach the air transport market with vigor and energy.[23] The preeminent consideration in the low cost-legacy carrier equation is that the low cost carrier is not a trend but rather a business model with imposing and permanent visibility in the market place. Faced with the prospect that low cost carriers are here to stay, legacy carriers (sometimes called full service network carriers) may serve a declining share of the market which could eventually stabilize at around 40–50 %[24] although with a higher revenue share. One has to consider that the threat posed by low cost carriers to the legacy carriers was non-existent until the early 1970s, when charter carriers commenced encroaching on the market. This trend, together with the increasing awareness, particularly in Europe where controls on market access, monopoly of air services by legacy carriers who were receiving State aid, and restrictions on pricing and frequency of services were perceived to be overwhelmingly anti-competitive and thus detrimental to the interests of the traveling public.[25] In the United States, the *Airline Deregulation Act* of 1978 also paved the way for liberalization of the domestic air transport market within the United States.

---

[19] Assembly Resolutions in Force (as of 5 October 2001) ICAO Doc. 9790at III-I.

[20] *Id.* III-9, Appendix G.

[21] Policy and Guidance Material On the Economic Regulation of International Air Transport, ICAO Doc 9587, Second Edition—1999, Part 4 (4-1 to 4-23).

[22] *Manual on the Regulation of International Air Transport*, Second Edition—2004, ICAO Doc 9626, Chapter 4.2 at 4.2-1.

[23] Lawton (2002), p. 1.

[24] Tretheway (2004) at p. 4.

[25] Doganis (1992) at p. 15.

The success of the low cost carrier lies mainly in much lower cost structures when compared to their larger competitors, efficient seat management policies and the absence of discrimination on price to the same extent as the legacy carriers. Particularly, the one way low air fare offered by the low cost carrier obviated the tedium of the Saturday stopover requirement for the traveler together with a drastically reduced air fare between two points, which more than compensated for the frills such as network connectivity and other value added services offered by the legacy carriers.

The secret of success of the low cost carrier is in the advantage it gains in being able to drastically lower its unit costs, unlike long haul carriers which are often burdened with a high cost base. Low cost carriers have, for several years, cut working capital and brought pressure on their asset bases, mostly by reducing their inventories and getting rid of uneven paying practices. This brings to bear the blatant difference between long haul legacy carriers, whose working capital is significant and low cost carriers whose asset base is comparatively low. For example, a decade ago, while the net operation of working capital of Alitalia was 8.1 % of its sales, and Air France retained a workable capital level of 2.2 % of sales, Ryanair in contrast showed a negative working capital of 2.8 % of its sales for the same year.[26]

Efficient working capital management is a competitive tool used by most low cost carriers in strengthening their balance sheets and credit rating. Good capital management also drives the cash flow and revenue—expenditure base of an airline which makes it possible to offer the customer a low priced but quality product. Thus the low cost carrier phenomenon is transcending the traditional boundary, from being a point-to-point basic service with the barest of minimums in frills, to becoming the provider of good service with hot and cold meals, beverage and bar service, video and audio entertainment, good seat pitch, blankets and pillows—all made possible by simplifying business processes. There has been a marked surge of energetic and robust competition among carriers in the air transport industry in recent years, mostly due to the globalization of the industry and privatization of airlines and airports. Regionally, market deregulation in Europe and Asia added to the impetus of emerging trends, collectively resulting in the emergence of a number of new price-based carriers, as well as the restructuring of existing carriers to meet the competition. The business models that have been applied by some major network airlines have proved to be fundamentally flawed in the past few years, enabling emergent low cost carriers to establish themselves with robust business profiles.[27] These low cost carriers have adopted sustained pricing policies

---

[26] Korfman and Overeynder (2003–2004) at p. 28.

[27] Doganis observes that in 1999 a survey carried out on 19,000 leisure passengers had astonishing results where the majority had preferred low cost no-frills carriers to established scheduled airlines such as British Airways and other scheduled European air carriers. See Doganis (2001) at p. 126. In 2008, a study conducted by a leading market research company Synovate revealed that the most popular response to a survey question "what is the best thing about air transport" was the answer

consistent with the recovery of costs and profit making. The end result has been the dwindling of market shares of legacy carriers. The main difficulty faced by legacy carriers was their inability to maintain a viable business model that could drive a revenue base to cover a traditional cost base while allowing for an adequate return on invested capital.

Inherent to the metamorphosis was the radical change from a conventional commercial profile which existed from 1944 (when regulated international civil aviation was introduced) to the end of the twentieth century where a conventional purchasing structure existed permitting the purchase of air travel, anywhere in the world—seamlessly, to a system of air travel accessible through the relatively expeditious and simple means such as e-mail and the internet. Also, in the past, a complex set of relationships existed, called interlining, which offered the customer an infrastructure and procedures to be connected to a network of airlines. Aided by standards and recommended practices of ICAO and proactive regulations of IATA, customers who wished to travel internationally by air enjoyed low transaction costs and comfortable travel, ensured through a single call to a travel agent. These transactions ensured flexibility, refundability of fares in case of failure to travel, and often transferability to interlined carriers and others who were participating in the network of agreements that prevailed.

7.3 A Sensible Approach to Fragility?

Even a cursory look at ICAO's role in the economic regulation suggests that the organization would do well to adopt an antifragile approach rather than the teleo-logical path it has taken all these long years, and continues to take. The Chicago Convention clearly identifies as one of ICAO's aims: "prevent economic waste caused by unreasonable competition". There is an identifiable degree of optionality in the word "prevent" calling ICAO to take the lead in guiding airlines and States on what they should be doing with their air services. Instead, the Council of ICAO takes a tepid stand and allows market forces to dictate, which inevitably results in the consumer being prevented of connectivity and value for money, which is ICAO's aim under Article 44 (d) of the Convention. ICAO does not know what economic waste there might be to prevent it. It would be an error for the Organization to think it knows where air transport is going, purely because the other players, i.e. the airlines, airports and service providers do not know where they are going themselves.

The most pragmatic way to approach this issue would be to apply the principles of the *Game Theory*. Otherwise known as the Interactive Decision Theory, it is a study of strategic decision making through a balance between conflict and

---

"Its fast and it gets me to where I need to be quickly". This is typically what a low fare, low cost carrier offers. See Clark (2010), p. 4 at p. 45.

cooperation between intelligent and rational decision makers. The air transport industry has, as the Game Theory ascribes, been subject to the strategic interactions among economic agents (air carriers), whose success has depended on the choices made by others (the States). Strategic alliances among carriers have sprung up; franchising is quite common and "deals" between the protagonists have been seen. The lack of global leadership is one thing; but the strangulation of the industry by regulatory overkill is another, and ICAO has to lead the way into a realm of cooperation, particularly with a close partnership with IATA, as earlier discussed.

ICAO has a glowing example of antifragility in its air navigation (or technical) field in the form of the Aviation System Block Upgrades (ASBUs)[28] which could well be an analogy to the economic regulation of air transport. ASBUs are quintessentially optional, each of which having a 5 year time scale. The ASBUs are not overarching, nor are they an umbrella system but remain flexible modules that can be used by States in accordance with their individual operational needs. One of the salient characteristics of ASBUs is that they define technologies and procedures that are calculated to improve operational performance, particularly when the need arises for an operational problem to be solved. ASBUs are based on flexibility and collaboration and as such are not mandatory requirements imposed on States The ultimate aim is to achieve global harmonization and interoperability of air navigation. The development of economic modules to assist States with their provision of air transport services could be modelled on the same basis, provided there is close cooperation and partnership between ICAO and the key partners, IATA, ACI and CANSO.

The basic philosophy of the ASBU is although there have been unparalleled advances in aviation safety into the twentieth century, this trend has to be sustained in order to keep pace with the rapid advances of aviation and the exponential growth of air transport.[29] The most effective way in which to keep pace and ensure a high rate of safety in air transport is by increasing the use of information that would identify potential safety problems beforehand. Critical to this equation is the gathering of information.[30] Some key areas in which the exchange of information could be successfully achieved are System-Wide Information Management (SWIM), Flight Flow, Digital AIM and ATM Information which would significantly improve service, and airborne separation and network operations, all which are specific areas addressed by the ICAO Aviation System Block Upgrades (ASBUs). Similarly, information gathering and developing economic models that would cater to the growing needs of the travelling population could provide a sound basis towards defragilizing the economics of air transport and making the industry flexible.

---

[28] See Abeyratne (1914), pp. 1–20.

[29] ICAO forecasts that, by the year 2030, scheduled passenger traffic around the world is expected to be more than double, from 2.9 billion in 2012 to over 6 billion passengers annually. See *Sustainability and Economic Development of Air Transport*, ATConf/6-WP/22 at 2.

[30] See Voss (2011), at p. 7.

The economic models should be viewed in the context of their main purpose—as modules that are developed and designed to achieve harmonization and interoperability that result in improvements in the provision of air transport services globally. Harmonization in this context is consistency in procedures and practice. Since not all these modules will be intended to be implemented everywhere simultaneously, they have to be prioritized in terms of their implementation to assist in the determination as to the circumstances, places and timeframes in which they should be implemented.

It is clear that the Council has no coercive power over States in their implementation of ASBUs. The Council, under the Convention, has only functions (which are in essence duties) and no powers.[31] On the other hand the Assembly has powers and duties accorded to it in the Chicago Convention,[32] one of which is to delegate to the Council the powers and authority necessary or desirable for the discharge of the duties of the Organization and revoke or modify the delegations of authority at any time.[33]

Article 54 (n) provides that the Council can consider any matter relating to the Convention which any Contracting State refers to it, giving the Council the capacity to make its own determination and recommendations pertaining to a matter referred to it. In this context, the Council may invoke Article 44 which identifies as one of the aims and objectives of ICAO fostering the development of air transport by meeting the needs of the people of the world for air transport.

A significant issue in the determination of ICAO's effectiveness as an international organization is the overriding principle of universality and global participation of all its 191 Contracting States in the implementation of ICAO policy. This principle, which has its genesis in the Chicago Conference of 1944, has flowed on, gaining express recognition of legal scholars. This is what makes ICAO unique as a specialized agency of the United Nations and establishes without any doubt that ICAO is not just a tool of cooperation among States.

Notwithstanding the Council's limited scope on the implementation of ASBUs, the Council could robustly monitor their implementation. Article 55 (c) requires the Council to:

---

[31] Although Jacob Schenkman, in his well documented and logically reasoned treatise on ICAO states that "The Council has been entrusted with duties, powers and functions..." he does not give a single example of such a power. See Schenkman (1955) at p. 158. For an exhaustive description and discussion of the ICAO Council, see Milde (2012), at pp. 149–180. Milde states: "The ICAO Council is a body of unique characteristics in the entire United Nations system of organizations. It possess not only the typical administrative and management functions within the organization but it is endowed with functions of a law making nature (frequently called "*quazi-legislative*") and functions in the settlement of differences (sometimes called "*quazi-judicial*"). *Id.* at 149. Weber, in his book gives a description of what he calls "a multitude of legislative, administrative and judicial functions" of the Council. See Weber (2007), pp. 21–26 at 22–23. See also, Abeyratne (2009), at pp. 278–285.

[32] Article 49 of the Chicago Convention.

[33] Article 49 (h) of the Chicago Convention.

Conduct research into all aspects of air transport and air navigation which are of international importance, communicate the results of its research to the Contracting States, and facilitate the exchange of information between Contracting States on air transport and air navigation matters.

This could be tied to the objective of ICAO to meet the needs of the people of the world for safe, regular, efficient and economical air transport. In this regard the Council should initiate studies that involve research into economics of air transport, which is of paramount international importance. This is provided for in the Chicago Convention.[34] Such studies, taking into account global, regional and national implementation trends, could analyse their effects on the improvements on safety regionally and globally. This in turn could result in a compendium of implementation of economic measures for States and airlines.

Such economic measures should be based on the current reality that ICAO's purpose in air transport is to focus on the user's interest (as prescribed by the Chicago Convention) rather than the air transport provider's choices. ICAO's aim should be not to pander to a hackneyed concept of sovereignty. Dr. Rudi Teitel, Professor of Comparative Law at New York Law School and Visiting Professor, London School of Economics, Global Governance, in her book *Humanity's Law*[35] quotes former United Nations Secretary General Kofi Annan who said:

State sovereignty, in its most basic sense, is being redefined – not least by the forces of globalization and international co-operation. States are now widely understood to be instruments at the service of their peoples and not vice versa. At the same time individual sovereignty,- by which I mean the fundamental freedom of each individual, enshrined in the Charter of the UN and subsequent international treaties – has been enhanced by a renewed and spreading consciousness of individual rights....[36]

Teitel follows with her own observation:

sovereignty is no longer a self-evident foundation for international law. This shift is driving the move from the State-centric normative discourse of global politics – which had prevailed until recently – to a far ranging, transnational discourse in which references to changed subjectivity have consequences. That new discourse is constructed more among humanity law lines[37]

This statement is consistent with the pronouncement of the International Criminal Tribunal for the former Yugoslavia which in its adjudication of *Prosecutor* v. *Dusko Tadic* said:

a state-sovereignty oriented approach has been gradually supplanted by a human being oriented approach.[38]

---

[34] Articles 55 (c) and (d).

[35] Oxford University Press: 2011.

[36] *Id.* at 9.

[37] *Id.* at 9–10.

[38] Decision on the Defence Motion for Interlocutory Appeal on Jurisdiction (2 October 1995), Appeals Chamber, International Criminal Tribunal for the Former Yugoslavia, paragraph 97.

Placing these views in an aviation context, perhaps the most important observation at ATConf/6 was made by Jeff Poole, Director General of CANSO when he called on States to help realise the vision of global seamless air traffic management (ATM) by employing a more proactive use of sovereignty over their airspace. At ATConf/6 it was CANSO's view that t States can legitimately delegate responsibility for the provision of air navigation services to another State or third party provider. Such delegation is a responsible use of sovereignty powers and would lead to much more effective, globally harmonised and cost-effective ATM. Some States have already exercised their sovereignty in this way to reap considerable benefits and CANSO urged others to follow their lead.

Poole, said:

> Air navigation services require a global, seamless, and performance-based approach to management of airspace, rather than one based on national borders. For too long States have misused the concept of airspace sovereignty as an ill-founded excuse to resist much-needed changes in ATM. Delegating responsibility for the provision of air navigation services does not mean that States give up their sovereignty or put national security at risk. We are simply asking States to join with other States to institute an air traffic management system that is not hampered by rigid and unnecessary adherence to national borders.
>
> Let's slay the sovereignty myth once and for all. Collaborating on cross-border arrangements will result in massive improvements in ATM performance. Sovereignty should be seen and used as an enabler of the changes required for more efficient management of the global air navigation system. We urge States to use their sovereignty powers to drive improvements in ATM performance and help to achieve the goal of seamless ATM globally. The benefits are huge – greater efficiency, fewer delays, lower CO_2 emissions and considerable cost savings.

CANSO further submitted that the delivery of cross-border services is fully compatible with the sovereignty of the airspace of States. Under the Chicago Convention, each State has complete and exclusive sovereignty over the airspace above its territory. It must provide air navigation facilities and services to facilitate international air navigation in its territory. However, this responsibility can be delegated—Article 28 of the Convention does not oblige States to provide air navigation services over their territory themselves. If a State decides to delegate, it still retains full power and authority in the airspace over its territory and prescribes the conditions under which the delegation is agreed. It must ensure that the service delivery activity is properly regulated; the designated service provider is certified; and adequate and effective supervision is exercised. The delegating State ensures there is a regulatory framework which establishes the overall performance standards for safety, efficiency and the environment. Service Level Agreements can be enacted that include Key Performance Indicators and targets which act as a powerful incentive to perform. Failure to meet the performance criteria can result in the delegation being revoked.

Most would agree that, in today's context, national security is not compromised by delegation. States need to monitor national airspace for security purposes, and to be able to respond to security or military threats. When a State delegates, it needs to implement appropriate arrangements and incentives to ensure that there is good military/civil cooperation. In the event of a crisis or threat to national security, the delegating State has the power to withdraw the delegation with immediate effect.

There are good examples of successful cross-border service provision across the world. There is mutual delegation between the USA and Canada; Tonga and Samoa delegate to New Zealand; there are various delegations in Europe from and to Finland, France, Norway, Sweden and Switzerland. Other States should look to these examples to see what can be achieved when artificial airspace boundaries are removed.

Thomas Jefferson once wrote that the purpose of government is to enable the people of a nation to live in safety and happiness. Government exists for the interests of the governed, not for the governors. As Benjamin Franklin wrote, "In free governments the rulers are the servants and the people their superiors and sovereigns." The ultimate powers in a society, therefore, rest in the people themselves, and they should exercise those powers, either directly or through representatives, in every way they are competent and that is practicable.

References

Abeyratne R (1914) ICAO's aviation system block upgrades – towards global harmonization and interoperability. Air Space Law XXXVIII(3):1–20

Abeyratne R (2009) Aeropolitics. Nova science Publishers, New York

Barrett SB (2004) How do the demands for airport services differ between full service carriers and low cost carriers? J Air Transport Manag 10(1):36

Clark P (2010) Stormy skies. Ashgate, Farnham

Doganis R (1992) The impact of liberalization on European airline strategies and operations. J Air Transport Manag 1(1):15

Doganis R (2001) The airline business in the 21st century. Routledge, London

Dunn G, Govindasamy S, Ranson L (2009) Asia leading low cost growth. Airline Bus VIII:56

Franke M (2004) Competition between network carriers and low cost carriers – retreat battle or breakthrough to a new level of efficiency? J Air Transport Manag 10(1):17

Gillen D (2004) Competitive advantage of low-cost carriers: some implications for airports. J Air Transport Manag 10(1):43

R Korfman, W Overeynder (2003–2004) Liberation of capital. Airlines Int, IATA, Issue 12/1:28

Lawton TC (2002) Cleared for take off, structure and strategy in the low fare airline business. Ashgate Publishing Limited, Aldershot, p 1

Milde M (2012) International Air Law and ICAO. In: Benko M (ed) Essential Air and Space Law, 2nd edn. Eleven International Publishing, The Hague

Samuelson P (1948) Economics. Mcgraw-Hill/Irwin, New York

Schenkman J (1955) International Civil Aviation Organization. Librairie E. Droz, Geneve

Taleb NN (2012) Antifragile. Random House, New York

Tretheway MW (2004) Distortions of airline revenues: why the network airline business model is broken. J Air Transport Manag 10(1):3–14

Voss B (2011) Aviation safety concern, air transport in the 21st century – key strategic developments. In: O'Connell JF, Williams G (eds) Ashgate

Weber L (2007) International Civil Aviation Organization – an introduction. Kluwer Law International, Dordrecht, pp 21–26

Chapter 8
Outcome of the Sixth Air Transport Conference (ATConf/6)

ATConf/6 was attended by 1,094 delegates from 132 Member States of ICAO, 49 observer delegations and 49 other delegations.

In its overview of the air transport industry, ATConf/6 suggested that ICAO follow up work relating to ATConf/5, in the field of economic regulation, facilitate and assist States in adapting to changes in the process of regulatory evolution. The continuation of such work would be beneficial to States in the development of air transport with a view to fostering sustainable development of air transport, distinguishing between profitability differences within the air transport value chain, avoiding fragmentation in the regulatory framework, and meeting challenges associated with financing the air transport system, infrastructure capacity constraints and unilateral actions by States that have an impact on international air transport. These are all impediments to sustainable development of air transport and eliminated in order to cope with the expected growth of world air traffic while fostering sustainable development of the air transport system. The Conference noted that the cost of fuel is largely dependent on market forces and that the possibility of regulatory intervention appears unlikely. However, it was imperative that the regulatory framework for the development of air transport foster strategies to mitigate the negative economic impact of fuel price volatility.

ATConf/6 also noted that, during the past decade, stakeholders of the air transport value chain had been operating in an increasingly competitive environment. These stakeholders were interdependent. Also, operating costs and the operational environment impact the value created by the air transport industry. Alliances, initially created to respond to market regulation restrictions and to remain competitive and present on the market, had become increasingly complex and interrelated in terms of cooperation frameworks and enormous marketing powers. Despite all the features and benefits of alliances and mergers in the air transport industry, there were drawbacks and flaws which need to be addressed and ICAO should conduct studies on this issue.

The Conference was of the view that both States and relevant international organizations should take a broad, homogeneous and sustainable approach to air transport and tourism policies, and, taking into consideration the different realities

R. Abeyratne, *Regulation of Air Transport*, DOI 10.1007/978-3-319-01041-0_8,
© Springer International Publishing Switzerland 2014

of States, recognized that liberalized aviation markets have generally produced significant growth, development and social benefits for States that had embraced them. There was a suggestion that, in order for ICAO to play an important role in air transport liberalization, a dedicated air transport fund may be established with a view to seeking voluntary contributions from Members States so as to enhance the work of the Organization in this field, provided that such a fund is administered in accordance with ICAO rules of governance and policies.

8.1   Discussions and Achievements of ATConf/6

The Conference recommended generally that States recognize the importance of national and regional regulatory frameworks in ensuring compliance of alliances with competition standards and in preventing monopolies and that States should also give due consideration to the benefits that alliances create. It was also stressed that States should consider the creation of mechanisms that allow for closer co-operation and co-ordination between their tourism and air transport authorities. ICAO was requested to continue to monitor developments, conduct studies on major issues of global importance, provide a set of basic principles to States and share its analyses on the development of the air transport industry with States, international organizations and the industry. Also ATConf/6 suggested that ICAO update and advance its guidance material on the regulation of international air transport. In particular, the Conference recommended that it should continue to update the Template Air Services Agreement (TASAs) to keep pace with regulatory evolution and to update liberalization indicators. ICAO should also continue to develop relevant databases such as the Database of the World's Air Services Agreements (Doc 9511), as well as case studies of liberalization experiences. Another task for ICAO was to continue to assist States with air transport liberaliza- tion efforts. This could be undertaken, *inter alia*, through the development of additional training courses, regional seminars or similar activities for the benefit of States, in accordance with available resources.

ICAO was recognized as the only forum for initiating global solutions for the development of a sustainable air transport system for all interested parties and was requested to continue to cooperate with international and regional organizations and with the industry in order to monitor impediments to a sustainable air transport system and define, in a cooperative manner, key strategies to overcome impediments as well as establishing an air transport fund in order to seek voluntary contributions from Member States with a view to enhancing the work of the Organization in this field. This fund should be administered transparently in accordance with relevant ICAO rules of governance and policies.

The Conference broadly addressed seven areas: market access; ownership and control of airlines; consumer protection; taxation; fare competition; safeguards in the liberalization process; and economics of airports and air navigation services. These areas were elaborated upon in an ICAO news release at the commencement of the Conference to the effect that the Conference will endeavour "to resolve pivotal matters relating to improved competition and market access through increased 'open skies' approaches, lingering state restrictions on airline ownership and control, consumer protection and price transparency for global passengers and the on-going proliferation of taxes and fees that are making it more expensive for people to travel by air".[1]

8.1.1 Market Access

There was general recognition of the broad benefits that liberalization of air transport provides to States, the industry and consumers, as well as its contribution to the global economy, trade, tourism, community development and job creation and the Conference expressed support for the need for further market access liberalization. The discussion focused on the approaches and pace of liberalization by States, as well as on the need to consider the interests of all stakeholders, including labour.

Many endorsed the view that due to the disparity in States' developmental stages and air transport requirements, States should pursue liberalization at their own pace and apply approaches suitable to their needs and national situation. At the same time, there was general agreement on the need to modernize the global regulatory framework on market access so as to adapt to the changes of a globalized business environment; also recognized was the need for ICAO to play a leadership role in facilitating regulatory evolution. In this regard support was voiced for the proposal that ICAO develop a long-term vision for global liberalization of air transport, including multilateral solutions, bearing in mind the interests of all States and aviation stakeholders.

Recognizing the distinct features of air cargo operations, the Conference voiced broad support for the proposal that ICAO develop an international agreement for the liberalization of air cargo transport. Pending such an agreement, States were encouraged to continue liberalization of air cargo services using existing vehicles, including bilateral, regional or multilateral approaches. Attention was drawn to the fact that States may join the Multilateral Agreement on the Liberalization of International Air Transportation (MALIAT) on a cargo-only basis. The Conference noted that the provisions of MALIAT, and the reasons for which more States have not joined, should be considered when developing the new agreement.

A number of States expressed support for the proposal that ICAO explore the development of an international agreement, building on past achievements and existing liberalized regulatory arrangements, for States to use in liberalizing market access. However, they noted that substantial work might be required in this regard.

---

[1] http://www.icao.int/Newsroom/Pages/once-a-decade-ICAO-air-transport-conference-convenes.aspx.

It was also suggested that this exercise be undertaken through consultation with experts (e.g., Air Transport Regulation Panel (ATRP)), States, the industry, and other aviation stakeholders, and that prior to committing to the project, ICAO should conduct a survey to analyse the needs of States, and assess the value, and resources required, for developing such an agreement.

In light of the lengthy process required for developing multilateral agreements, there was wide-spread support for the proposal that States should continue to liberalize market access through existing avenues, including bilateral, regional and plurilateral arrangements The Conference reached the following conclusions:

(a) Liberalization has provided broad benefits for States, consumers, airlines, airports, communities and national economies. Increased access to the international market for air service providers allows the air transport sector to maximize its contribution to the global economy;

(b) Liberalization is a means and process, not an end. The objective of regulatory evolution is to create a favourable environment in which international air transport may develop and flourish in an orderly, efficient, economical and sustainable manner, without compromising safety and security, while respecting social and labour standards;

(c) States have pursued liberalization of market access in various ways and at different paces according to needs, conditions and policy goals, and have achieved considerable progress and success. However, there is an increasing need to modernize the decades-old regulatory regime to adapt to a changed global economic and business environment and to meet the requirements of States, the industry and consumers in the twenty-first century;

(d) There is strong endorsement from the aviation community for further liberalization of air transport, and for ICAO's engagement in facilitating such efforts. There is broad support for ICAO, working closely with all parties concerned, to take the lead in exploring ways by which to expand market access, including long-term multilateral solutions, recognizing that this will be a continuing process; and

(e) In the short term, States should continue to pursue market access liberalization according to situations and requirements, using existing avenues, while ICAO should continue to provide guidance and assistance to States in facilitating the process.

On the above basis, the Conference recommended that:

(a) States should continue to pursue liberalization of market access at a pace and in a manner appropriate to needs and circumstances, giving due regard to the interests of all stakeholders, the changing business environment and infrastructure requirements;

(b) ICAO should develop and adopt a long-term vision for international air transport liberalization, including examination of the possibility of an international agreement by which States could liberalize market access, taking into account the existing experience and achievements of States, as well as the various proposals presented during the Conference;

(c) ICAO should work with all parties concerned, undertaking consultation with experts, States, the industry, interested organizations and other stakeholders to build a common understanding and obtain consensus for the development of the long-term vision and related regulatory arrangements;

(d) ICAO should continue to provide guidance and assistance to States in facilitating market access liberalization, using facilities such as the ICAO Air Services Negotiation Conference (ICAN); and

(e) ICAO should keep its policy guidance on air transport regulation and liberalization current and responsive to changes and to the needs of States, and consider additional means by which to facilitate liberalization.

It is noteworthy that, after ATConf/5 in 2003, not much seems to have changed in the field of liberalization. For example, ATConf/6 in its Report as stated in (a) above says: "States should continue to pursue liberalization of market access at a pace and in a manner appropriate to needs and circumstances, giving due regard to the interests of all stakeholders, the changing business environment and infrastructure requirements". The ATConf/5 Report said much the same thing—that: "States should continue to pursue liberalization in this regard at their own choice and pace..."[2] ATConf/6 also recommended that: "ICAO should keep its policy guidance on air transport regulation and liberalization current and responsive to changes and to the needs of States, and consider additional means by which to facilitate liberalization". This is a reiteration of the ATConf/5 Report: "...ICAO should continue to keep current the existing guidance material on the economic regulation of international air transport..."

A notable deviation in ATConf/6 from its predecessor of 10 years earlier was the absence of any reference to bringing market access under the General Agreement on Trade in Services (GATS) and, surprisingly the absence of mention by ATConf/6 of the International Air Services Transit Agreement (IASTA) which ATConf/5 alluded to. The only significant point that seemed to emerge in this extremely important issue was that ICAO should develop and adopt a long-term vision for international air transport liberalization, including examination of the possibility of an international agreement by which States could liberalize market access, taking into account the existing experience and achievements of States, as well as the various proposals presented during the Conference.

One does not need a global conference to come to this conclusion, and *a fortiori*, ICAO does not need to wait for an entire conference to be told to take this initiative. In the least the Conference should have discussed and outlined the key issues in such a vision; the issue of regional liberalization versus global liberalization; and the timelines involved. As it stands, ICAO has 10 years to develop that vision!!

---

[2] *Report of the Worldwide Air Transport Conference, Challenges and Opportunities of Liberalization* (Montreal, 24–28 March 2003) Doc 9819, ATConf/5 2003 at 28.

In terms of air cargo liberalization ATConf/6 concluded:

(a) Air cargo plays an important role in the global economy. The growth and expansion of air cargo services is beneficial for the sustainable development of air transport, and contributes significantly to global trade and economic development;

(b) The distinct features of air cargo services need to be given due consideration by States when making air service arrangements;

(c) States have used various vehicles in liberalizing air cargo services, including bilateral, regional and plurilateral arrangements, some of which are open for other States to join, such as the Multilateral Agreement on the Liberalization of International Air Transportation (MALIAT); and

(d) ICAO guidance on liberalization of air cargo services remains relevant, and its use by States should be encouraged. In this regard, there is broad support for ICAO to play a leadership role in facilitating further liberalization and for ICAO to develop a multilateral agreement specifically focussed on air cargo. The Recommendations of ATConf/6 were:

- States should give due regard to the distinct features of air cargo services when exchanging market access rights in the framework of air service agreements and grant appropriate rights and operational flexibility so as to promote the development of these services;
- States should continue to liberalize air cargo services through all available avenues, and to share experiences with other States;
- ICAO should initiate work for the development of a specific international agreement to facilitate further liberalization of air cargo services, taking into account past experiences and achievements, views of States on existing arrangements, and suggestions made during the Conference; and
- In the development of new regulatory arrangements on air cargo, ICAO should engage all parties concerned, and should undertake consultation with experts, States, the industry and interested stakeholders.

Surprisingly, the recommendations of ATConf/6 on air cargo liberalization are much less effective and certainly less forward thinking than those of ATConf/5. For instance, while ATConf/6 recommends in a general manner that States should continue to liberalize air cargo services through all available avenues, and to share experiences with other States, ATConf/5 more specifically and forcefully recommended: "States should use one or more of three progressive alternatives : unilateral liberalization of market access for all cargo services without bilateral reciprocity or negotiation; liberalizing all cargo services through bilateral arrangements and negotiations to ensure reciprocity; and using a multilateral/ plurilateral approach for the liberalization of all cargo services".[3]

---

[3] *Id.* at 32.

8.1.1.1 MALIAT

New Zealand suggested to ATConf/6 that MALIAT would be a good starting point towards multilateral liberalization. MALIAT[4] was negotiated on 31 October to 2 November 2000 at Kona, Hawaii, and signed at Washington D.C. on 1 May 2001 by Brunei Darussalam, Chile, New Zealand, Singapore and the United States of America. It entered into force on 21 December 2001. The key features of MALIAT are :an open route schedule; open traffic rights including seventh freedom cargo services; open capacity; airline investment provisions which focus on effective control and principal place of business, but protect against flag of convenience carriers; multiple airline designation; third-country code-sharing; and a minimal tariff filing regime.

MALIAT grants first to fifth freedom rights (inclusive) to States Parties who are signatories as well as "rights otherwise specified in the Agreement".[5] They may operate flights in either or both directions; combine different flight numbers within one aircraft operation; serve behind, intermediate, and beyond points and points in the territories of the Parties on the routes in any combination and in any order; omit stops at any point or points; transfer traffic from any of its aircraft to any of its other aircraft at any point on the routes; serve points behind any point in its territory with or without change of aircraft or flight number and hold out and advertise such services to the public as through services; make stopovers at any points whether within or outside the territory of any Party; carry transit traffic through any other Party's territory; and combine traffic on the same aircraft regardless of where such traffic originates.

MALIAT also provides that on any segment or segments of the routes above, any designated airline may perform international air transportation without any limitation as to change, at any point on the route, in type or number of aircraft operated; provided that, with the exception of all-cargo services, in the outbound direction, the transportation beyond such point is a continuation of the transportation from the territory of the Party that has designated the airline and, in the inbound direction, the transportation to the territory of the Party that has designated the airline is a continuation of the transportation from beyond such point.

MALIAT does not permit cabotage.[6] It also has an ownership and control clause similar to the traditional clause in bilateral air services agreements. MALIAT focuses heavily on security, where, in accordance with their rights and obligations

---

[4] The Agreement is open to accession by any state that is party to the following aviation security conventions: the *Convention on Offences and Certain other Acts Committed on Board Aircraft*, done at Tokyo on 14 September 1963; the *Convention for the Suppression of Unlawful Seizure of Aircraft*, done at The Hague on 16 December 1970; the *Convention for the Suppression of Unlawful Acts against the Safety of Civil Aviation*, done at Montreal on 23 September 1971; and the *Protocol for the Suppression of Unlawful Acts of Violence at Airports Serving International Civil Aviation*, done at Montreal on 24 February 1988.

[5] Article 2.1.

[6] Article 2.5.

under international law, the State Parties to the Agreement are required to reaffirm that their obligation to each other to protect the security of civil aviation against acts of unlawful interference forms an integral part of the Agreement. Without limiting the generality of their rights and obligations under international law, the Parties are required in particular act in conformity with the provisions of the Convention on Offenses and Certain Other Acts Committed on Board Aircraft, done at Tokyo on September 14, 1963, the Convention for the Suppression of Unlawful Seizure of Aircraft, done at The Hague on December 16, 1970, the Convention for the Suppression of Unlawful Acts against the Safety of Civil Aviation, done at Montreal on September 23, 1971, and the Protocol for the Suppression of Unlawful Acts of Violence at Airports Serving International Civil Aviation, done at Montreal on February 24, 1988.

Recommendation (c) of ATConf/6, referred to above, hints at the relevance of MALIAT when ICAO attempts at developing a multilateral agreement. Although this would be a positive step, ICAO should also focus on the importance of connectivity from the point of view of offering the best deal in terms of the least connecting time at a reasonable and accessible price, which is the cornerstone of the Chicago Convention as enunciated in Article 44 (d).

8.1.2   Ownership and Control

At ATConf/6, there was wide recognition of the benefits of liberalizing air carrier ownership and control; also noted was the need to adapt the current regulatory regime to meet the needs of the twenty-first century. There was general endorsement that ICAO is the appropriate forum by which to promote and facilitate further liberalization in this area.

There was virtually unanimous support for ICAO to take the lead and, as a matter of priority, initiate work to develop an international agreement to facilitate liberalization of air carrier ownership and control, building on past achievements and experiences. There was broad agreement that such work should take into consideration the importance of ensuring safety and security, the principle of reciprocity, and the need for safeguards to allow a gradual and progressive adaptation. It was acknowledged that this work should also take account of regional liberalization experiences, the rules under various States' domestic laws, and the effects on all stakeholders, including labour.

The Conference noted the distinction drawn between a State's decision with respect to the ownership and control requirement of its national airlines under domestic law, and the discretion of a State to include or remove restrictions on ownership and control of designated airlines of other States under bilateral air service agreements.

There was also broad support for encouraging States to continue to liberalize air carrier ownership and control through various existing measures, including that of adopting the alternative criteria for designation of airlines based on "principal place of business and effective regulatory control", as recommended by ICAO. In addition, many States expressed support for ICAO work required to update guidance in this area as necessary. The Conclusions of ATConf/6 on ownership and control of airlines were:

(a) Since ATConf/5, States and the industry have continuously called for liberalization of air carrier ownership and control in order to adapt to an ever more global and competitive economic environment. Although diverging views and regulatory approaches remain, there has been increasing recognition of the benefits of liberalization;
(b) More States are willing to liberalize and adopt a flexible approach in dealing with airline designation requirements under bilateral air service agreements. Considerable progress has also been achieved at the regional level, as several regions or sub-regions adopted arrangements in liberalizing air carrier ownership and control among their respective members, including permitting cross-border investments and airline mergers;
(c) There was general agreement that each State may choose its own path and pace in liberalization, and that safety and security remain of paramount importance in any regulatory change. Due consideration should also be given to the interest of all stakeholders, including that of labor. States have used various means in the liberalization of air carrier ownership and control, including unilateral, bilateral, regional, plurilateral and multilateral approaches;
(d) ICAO has developed considerable guidance for use by States in liberalization of air carrier ownership and control, which needs to be kept current and responsive to changes and to the requirements of States;
(e) There is broad and strong support for ICAO to play a leadership role in facilitating further liberalization and in modernizing the global regulatory framework, including the development of a multilateral agreement in order to meet the changing economic environment and requirements of States for the efficient and sustainable development of air transport; and
(f) ICAO should work with concerned parties and aviation stakeholders in developing new regulatory frameworks. Before a new long-term regulatory regime is in place, States could consider some short-term options proposed during the Conference for liberalization.

Along these lines ATConf/6 recommended as follows:

(a) States should continue to liberalize air carrier ownership and control, according to needs and circumstances, through various existing measures, such as waiver of ownership and control restrictions in bilateral air services agreements, and those recommended by ICAO. Regional organizations should, in cooperation with ICAO, play a role in facilitating and assisting States in the liberalization process;
(b) ICAO should continue to promote its policy guidance on air carrier ownership and control and encourage States to use its guidance in regulatory practice. It should keep its policy guidance current and responsive to changing situations

and to the requirements of States; where required, ICAO should study and develop guidance on important issues that may arise as liberalization progresses;

(c) ICAO should initiate work on the development of an international agreement to liberalize air carrier ownership and control, taking into consideration safety and security concerns, the principle of reciprocity, the need to allow a gradual and progressive adaptation with safeguards, the need to take account of regional experiences, the requirements of various States' domestic laws, and the effects on all stakeholders, including labour;

(d) ICAO should involve all parties concerned in the development of the international agreement, and should undertake consultation with experts, States, aviation stakeholders and interested Organizations.

Again, what stands out is the suggestion that ICAO should "ICAO should, as a priority, initiate work on the development of an international agreement to liberalize air carrier ownership and control". If ATConf/6 were more proactive it should have concluded and recommended that ICAO must develop such an agreement, and not merely initiate.

It is indeed a pity that every once in 10 years, States assemble in all their pomp and circumstance to "discuss" the key issues pertaining to the economics of air transport and conclude that this area of air transport is not worthy of leadership of ICAO nor is it worthy of clearing the clouds of State sovereignty. The Air Transport Conference has, over the past three decades, become a mere celebration and has stopped at that. Even more ironically, at a pre ATConf/6 Symposium held at ICAO a day before the Conference convened (on 17 March 2003) a packed hall spoke of nothing but ICAO's leadership in addressing key issues such as market access and ownership and control of airlines. ICAO member States would do so earnestly and openly as long as their statements are given informally and off the record.

The sclerotic ownership and control obstacle that plagues the airline industry and effectively precludes its growth has, at its core the blocking by States of foreign direct investment (FDI)—a subject I have dealt with in some detail in a chapter in one of my previous books.[7] The restriction imposed by the requirement that airlines be substantially owned and effectively controlled by nationals of the State of registry of the aircraft. This restriction, which effectively precludes FDI in air transport is the natural by product of sovereignty over the airspace above a State's territory, and presents a fundamental flaw in the economics of air transport and veers from the normative foundation of trade and development which is foreign direct investment (FDI).

A perceived anomaly in the requirement of substantial ownership and effective control is that, although States have liberally used the two terms in their bilateral air services agreements, these terms have not been formally defined in any modern air law instrument. They remain a conduit to State practice which pays lip service to

---

[7] Abeyratne (2012), at pp. 19–27.

the two terms with monotonous regularity and are treated as a "catch all" protection against FDI. The protection is calculated to protect national carriers from losing to foreign carriers what they believe is their "market share" of traffic. In the absence of definitive certainty the term "ownership" can only be surmised as more than 50 % ownership of company shares. Ownership legally defined is the exclusive legal title coupled with the legal right to possession, enjoyment of fruits and alienation of property. It is a collection of rights to use and enjoy property including the right to transmit it to others.

A majority ownership can be considered substantial. As for "effective control" the law contains no objective standard for what constitutes effective control of an airline. Thus regulators are free to interpret this concept according to national interests. Parties to the International Air Services Transit Agreement (IASTA) grant overflight rights for scheduled air services to an "air transport enterprise" that is substantially owned and effectively controlled by nationals of a Contracting States to IASTA. Article 1 (5) of IASTA stipulates that each contracting State reserves the right to withhold or revoke a certificate or permit to an air transport enterprise of another State in any case where it is not satisfied that substantial ownership and effective control are vested in nationals of a contracting State, or in case of failure of such air transport enterprise to comply with the laws of the State over which it operates, or to perform its obligations under this Agreement.

Defining "effective control" is more difficult than defining "substantial owner-ship" because, while ownership is usually transparent and can often be determined by public or other records of shareholders, effective control may be exercised in numerous different ways, many of which may not be transparent or readily perceivable. An example offered is that while air carrier management may exercise control over flight operations and other operations of an airline, its shareholders may exercise control over the injection of capital or dissolving the company.

The ownership and control requirement has upended the meaning and purpose of the Chicago Convention which makes it part of its philosophy in the *Preamble,* that international air transport services may be established on the basis of equality of opportunity and operated soundly and economically. *A fortiori*, if these services are to meet the needs of the people of the world for regular, efficient and economical air transport, such restrictions on international investment in the industry could be nothing but counter-intuitive.

It is by no means suggested that the ownership and control approach should be totally abandoned. Neither is it contended that the concept should be rigidly enforced to the detriment of the ultimate consumer. States are not entrepreneurs. At best, they could macro-manage their economies through regulations. Airlines on the other hand are entrepreneurs and should have the flexibility of flowing along market trends. The concept of sovereignty should be applied to issues of State and should not be used as a tool to control markets.

8.1.3 Consumer Protection

Consumer protection was another major subject under discussion at ATConf/6. The Conference expressed unanimous support for the need to protect consumers of air transport services. It was also noted that the effectiveness of regulatory responses adopted by States would benefit from increased convergence and compatibility.

There was broad support for the view that ICAO, as the best forum for addressing matters relating to international air transport, should play a leadership role in the development of high-level, non-prescriptive core principles on consumer protection. It was recognized that, in developing such core principles, it will be necessary to seek an appropriate balance between the need for consumer protection and industry competitiveness. It was also noted that the development of such core principles must allow flexibility with respect to State implementation, taking into account specific social, political and economic characteristics. Educating consumers on key aspects of air travel was considered an important aspect of consumer protection.

Several States and organizations noted that the development of core principles on consumer protection should be in line with existing instruments, notably the Convention for the Unification of Certain Rules for International Carriage by Air, adopted in Montréal on 28 May 1999 (i.e. the "Montréal Convention"). In this regard, States which have not done so were encouraged to adopt the Montréal Convention.

Regarding the development of the core principles on consumer protection in an efficient and expedient manner, support was expressed for the creation of a dedicated ad hoc group, which could emanate from existing ICAO bodies such as the Air Transport Regulation Panel (ATRP). In view of the benefits of providing users of air transport services with clear information on prices, there was support for the need to include price transparency as part of the core principles on consumer protection.

States supported measures aimed at enhancing air transport connectivity and endorsed cooperation in the identification of impediments thereto. With respect to further work in this area to be undertaken by ICAO, particularly that focused on cost-benefit analyses, it was suggested that further consideration is required, including that of the Aviation Security Panel (AVSECP) and the Facilitation Panel (FALP).

The World Tourism Organization (UNWTO) informed the Conference of its decision to postpone the development of its draft convention on the protection of tourists pending review of the outcomes of the Conference. The Conference agreed on the need to avoid the application of different sets of rules, based on whether a person is considered an air passenger or a tourist, notably in cases of massive disruptions or instances of "force majeure". Along the lines of these discussions ATConf/6 reached the following conclusions:

(a) The importance of protecting the interests of consumers is universally supported, as is the need for convergence and compatibility;
(b) Core principles on consumer protection, covering such issues as price transparency and assistance to passengers in the event of delays, cancellations and denied boarding, as well as persons with disabilities, would benefit both passengers and operators;
(c) States should be encouraged to become parties to the Convention for the Unification of Certain Rules for International Carriage by Air, adopted in Montréal on 28 May 1999;
(d) States should enhance air transport connectivity and work to identify and remove impediments thereto; States should ensure that passenger data and information available to government authorities, airlines, airports and other relevant sources are provided to ICAO to facilitate these efforts; and
(e) There is need for a coordinated approach amongst organizations, including the United Nations World Tourism Organization (UNWTO).

The Conference adopted the following recommendations:

(a) ICAO should continue to monitor consumer protection developments and to play a leadership role in developing policy guidance, taking into account the interests of States, the industry, air travellers and other aviation stakeholders;
(b) ICAO should, in particular, develop, in the short term, a set of high-level non-prescriptive core principles on consumer protection which strike an appropriate balance between protection of consumers and industry competitiveness and which take into account the needs of States for flexibility, given different State social, political and economic characteristics; these core principles should be consistent with existing instruments, in particular the Convention for the Unification of Certain Rules for International Carriage by Air, adopted in Montréal on 28 May 1999;
(c) ICAO should establish a dedicated ad hoc group drawn from existing bodies such as the Air Transport Regulation Panel (ATRP), including experts designated at ICAO's invitation by States or regional bodies, with a view to facilitating the development of the core principles in an efficient and expedient manner;
(d) ICAO should continue to play a leadership role in consumer protection in air transport and should cooperate with other international organizations, including UNWTO, in areas of common interest with a view to, inter alia, avoiding duplication of efforts;
(e) States should foster the adoption and implementation of consumer protection measures aimed at increasing the connectivity provided by air transport; and
(f) ICAO should take necessary action, possibly through the involvement of adequate bodies such as the Aviation Security Panel (AVSECP) and the Facilitation Panel (FALP), for subsequent work on cost-benefit analysis related to air transport connectivity.

The issue of consumer protection (or, as I would like to call it, consumer rights) was addressed by ATConf/5 under the heading "consumer interests".[8] The discussion therein was not as exhaustive as in ATCOnf/6 and ATConf/5 concluded that States needed to carefully examine what elements of consumer interests in service quality have adequately been dealt with by existing commercial practices of airlines (and service providers) and what elements needed to be handled by regulatory approaches or voluntary means. ATConf/5 also concluded that States needed to strike the right balance between voluntary commitments and regulatory measures, whenever government intervention was considered necessary to improve service quality. States were also requested to minimize unnecessary differences in the content and application of regulations. ICAO was requested to continue to monitor developments regarding voluntary commitments to and government regulation of consumer interests with a view to providing useful information to States to assist in the harmonization process.

ATConf/6 seemingly went a step further (albeit not a significant one) by ascribing to ICAO a "leadership" role in the development of guidance material and a set of core principles on consumer protection that would strike an appropriate balance between protection of consumers and industry competitiveness and that which take into account the needs of States for flexibility, given different State social, political and economic characteristics.

8.1.4 Fair Competition

At ATConf/6 ICAO advised that the reduction by States of controls within the air transport industry, known as "liberalization" or "deregulation", has fostered competition between air carriers. Enhanced competition, in turn, has led many carriers to consider consolidation as a means by which to achieve economies of scale and scope and to respond to consumer demands for global networks. The three major airline alliances, *Star Alliance*, *SkyTeam* and *Oneworld*, now represent more than 60 % of the global market share, measured in available seat-kilometres for total scheduled passengers. Competition today is not just between individual airlines but increasingly between these alliances. With heightened competition and consolidation has come a heightened risk of anti-competitive behaviour, including abuse of a dominant position and oligopoly practices. In addition, in order to keep their national airlines competitive in a liberalized market, some governments may be tempted to lend support to their airlines through means that could deny the airlines of other States a fair and equal opportunity to compete.

---

[8] *Report of the Worldwide Air Transport Conference, Challenges and Opportunities of Liberalization, supra* note 2 at 41–43.

The concept of fair competition is epitomised in the European approach where the EU air transport market has been the basis for further integration at a wider pan-European level. The "Common Aviation Area" being established through comprehensive aviation agreements between the EU and its neighbours is based on the same principles of open markets, regulatory harmonisation and the application of compatible competition rules to ensure fair competition.

The EU said at ATConf/6:

> Over time, air transport has been fully liberalised within the European Union (EU). Liberalisation has been progressing in parallel with regulatory harmonisation in all areas of civil aviation, which together has created a completely open, deeply integrated single air transport market where competition and State aid rules, as well as economic and technical requirements, are enforced by independent authorities at national and/or EU level. By supporting 5.1 million jobs and contributing €365 billion or 2.4% to European GDP3, the EU aviation sector makes a vital contribution to economic growth, employment, tourism, people-to-people contacts, as well as to regional and social cohesion. These significant benefits would not have been possible to gain within a framework which heavily restricts competition such as restricted bilateral air services agreements.[9]

As the global air transport market is becoming more and more open and competitive, it is more important to ensure that competition is not distorted by unfair practices. And where fair competition conditions exist, it is best to repeal or reduce market access restrictions e.g. in bilateral air service agreements so that airlines can compete freely. Indeed, fair competition is an important principle to achieve the objective of full liberalisation of market access and to reap its benefits. Furthermore, the existence of fair competition is also likely to encourage States to make further progress on liberalising airline ownership and control.

At ATConf/6 States recognized that the principle of fair and equal opportunity is enshrined in the Chicago Convention where States have agreed that international air transport services "may be established on the basis of equality of opportunity" and every State "has a fair opportunity to operate international airlines". What the Conference did not do was to interpret the words "a fair opportunity to operate international airlines". A fair opportunity is not a guaranteed right to operate air services to the detriment of fair competition.

A broad range of issues were raised under the topic of fair competition. A number of States focused on the challenges faced by smaller airlines, especially those from developing countries,when competing against much larger carriers, a challenge made more difficult in some cases by airline mergers and alliances as well as by unilateral or discriminatory measures that deny equitable opportunities. Others focused on issues linked to the inconsistent application of competition laws and policies, including standards for granting antitrust immunity. Some States noted the negative effects on competition caused by barriers to market access. Other States stressed that market liberalization must go hand-in-hand with concrete measures to ensure fair competition.

---

[9] Basic Principles of Fair Competition, *ATConf/6-WP/51*, at 2.

A number of States supported work by ICAO to establish core principles on fair competition, both to provide a clearer understanding of what is fair and unfair and to indicate appropriate measures to address problems. Among the measures identified were the establishment and effective enforcement of competition laws applicable to international air transport, clear and strong rules on state aids, and the inclusion of appropriate fair competition clauses in air services agreements based on ICAO templates. Other States mentioned principles of fair competition aimed at blocking control of markets by dominant carriers, ensuring all carriers equitable access, prohibiting discrimination and barring abusive practices. However, a number of other States disagreed with the proposal to establish core principles citing (a) that attempts to reach consensus on core principles would prove impossible given the major differences in State views and practices; (b) that issues of fairness can be effectively handled in existing bilateral channels and through use of ICAO template language; (c) that many airlines and airports are State-owned (hence core principles that challenge this fact would be contrary to the principle of State sovereignty in the Chicago Convention); and (d) that core principles on fair competition might be misinterpreted or misused as a barrier to competition.

The Conclusions of the Conference were that in accordance with the Chicago Convention, fair competition is an important general principle in the operation of international air services, and that ICAO policies on competition are still valid, based on observed practices, such as the inclusion of ICAO model clauses on competition in air services agreements. The Conference suggested that ICAO should continue to monitor developments and update its policies and guidance in response to changes in the industry and State practice, and that there is a recognized need for States to give due consideration to the concerns of other States in the application of national or regional competition laws and policies to international air transport as well as a need for cooperation among competition authorities, including in the context of approval of alliances and mergers. In this regard, the Conference was of the view that ICAO should play a leadership role in identifying and developing tools to promote dialogue and the exchange of information among interested authorities with the goal of fostering more compatible regulatory approaches. Such tools could include the development by ICAO of a detailed compendium of national and regional competition policies and practices as well as the development of a facility that would serve as a forum for the enhancement of cooperation, dialogue and exchange of information.

Accordingly ATConf/6 recommended as follows:

(a) States should take into consideration that fair competition is an important general principle in the operation of international air services;
(b) States, taking into account national sovereignty, should develop competition laws and policies that apply to air transport. In doing so, States should consider ICAO guidance on competition;
(c) States should give due consideration to the concerns of other States in the application of national and/or regional competition laws or policies to international air transport;

(d) States should give due regard to ICAO guidance in Air Services Agreements (ASAs) and national or regional competition rules;

(e) States should encourage cooperation among national and/or regional competition authorities, including in the context of approval of alliances and mergers;

(f) ICAO should develop tools such as an exchange forum to enhance cooperation, dialogue and exchange of information between Member States to promote more compatible regulatory approaches toward international air transport;

(g) ICAO should develop a compendium of competition policies and practices in force nationally or regionally; and

(h) ICAO should continue to monitor developments in the area of competition in international air transport and update, as necessary, its policies and guidance on fair competition through the Air Transport Regulation Panel (ATRP).

The only recommendation that stands out as being different on a general basis from those in ATConf/5 is the suggestion that ICAO develop a compendium of competition policies and practices in force nationally or regionally. In developing these policies, ICAO could do well to take into consideration the fact that fair competition is a key principle to achieve the benefits of liberalisation of market access in international air transport at worldwide level and that States should take measures to ensure fair competition, for example, through laying down efficient competition laws applicable to international air transport, as well as clear, transparent and strict state aid rules, developing and inserting fair competition clauses in their bilateral air services agreements, and through closer cooperation between their respective authorities including, where appropriate, in competition investigations. Critical to ICAO's considerations would be the possibility of introducing a fair competition clause to be included in bilateral air services agreements) to establish and maintain it at global level.

8.1.5 Safeguards in the Liberalization Process

There was general agreement with respect to the continued need for safeguards in the liberalization process in order to ensure the effective and sustained participation of all States in the international air transport system, as enshrined in the Chicago Convention. Many States considered ICAO guidance in this field to be useful for policy making and for the development of their air transport industry.

There was virtually unanimous support expressed with respect to the prevailing relevance and validity of ICAO's guidance on safeguard measures concerning participation, assurance of service and State aid/subsidies, essential air service and tourism development route scheme, and avoidance of unilateral action, as well as for the need for ICAO to keep its guidance current and responsive to the changes and requirements of States.

It was widely agreed that in regulatory practices, States should follow the ICAO policy guidance on safeguards according to their needs and situation. Particular emphasis was placed on the view that States should refrain from taking unilateral action that would negatively affect other States and the orderly, efficient and sustainable development of air transport.

The Conference noted the view that safeguard measures should not be used as a tool to hinder the liberalization progress. Noted also was a suggestion that ICAO should work with States, interested organizations and stakeholders to promote its guidance, and should consult States concerned when developing guidance on safeguard measures. The Conference concluded that:

(a) In the liberalization process of international air transport, there is a continued need for safeguards by some States due to the disparity in the stages of the development, strength of air carriers, and geographical locations, as well as the need to ensure sustainable development;

(b) The guidance developed by ICAO on safeguard measures pertaining to effective participation in international air transport, assurance of service and State aid/subsidies, essential air services, and avoidance of unilateral action, continues to be relevant, and should be kept current and responsive to changes and States' requirements; and

(c) In regulatory practices, States should give due regard to the common interest of the aviation community and the concerns of other States. Particular attention should be given to the ICAO policy guidance on the avoidance of unilateral action that could negatively affect the efficient and sustainable development of international air transport.

The Conference recommended that:

(a) In the liberalization process, States should give due regard to the principles agreed upon by the aviation community at the various ICAO fora pertaining to safeguard measures designed to ensure the sustained and effective participation of all States in international air transport, including the principle of giving special consideration to the interests and needs of developing countries;

(b) In regulatory practices, States should refrain from taking unilateral action that would negatively affect the common interest of the aviation community and the efficient and sustainable development of international air transport;

(c) ICAO should actively promote and encourage States to use the relevant ICAO guidance on safeguard measures in their regulatory practices, and to share with ICAO and other States their experiences in liberalization; and

(d) ICAO should continue to monitor developments with respect to safeguards, and should keep related guidance current and responsive to changes and needs of States and, where required, work with States, interested organizations and aviation stakeholders to develop further guidance.

There is a fundamental inconsistency in the principles recommended by ATConf/6 in that, if developing States are to be safeguarded to the extent that they are guaranteed "effective participation" this would inevitably erode the basic principle of competition which the Conference supported and also deprive the consumer of accessibility to the best possible product. This retrogressive approach revives the hackneyed "market share" argument of the seventies and eighties. In fact, we seem to have gone back to ATConf/4 of 1994 where one of the proposed future regulatory arrangements at the Conference was that parties would grant each other full market access (unrestricted route, operational and traffic) rights for use by designated air carriers, with cabotage and so called seventh freedom rights exchanges optionally. Of course, it was intended that each party would have the right to impose a time limited capacity freeze as an extraordinary measure and in response to a rapid and significant decline in that party's participation in a country pair market. The latter measure, called the "safety net" was intended to form a buffer against a total swing towards favouring unregulated commercial operations of air carriers. The market access and "safety net" principle was designed to award to each party's air carrier unrestricted basic market access rights to the other party's territories for services touching the territories of both parties (to the exclusion of cabotage rights, i.e. rights to operate commercial air services within points in the territory of another party) optionally, for so called seventh freedom services (i.e. services touching the territory of the granting party without touching the territory of the designating party); and/or optionally, with cabotage rights. To these rights, the "safety net" brought in the *caveat* that each party would have the right to impose a capacity freeze as an extraordinary measure, under six conditions that called for such a freeze. They were:

(a) To be implemented only in response to a rapid and significant decline in that party's participation in a country pair market;
(b) To be applied to all scheduled and non-scheduled fights by the air carriers of each party and any third State which directly serve the affected country-pair market;
(c) To be intended to last for a maximum finite period of, for example, 1 year, 2 years or 1 year, renewable once;
(d) To require close monitoring by the parties to enable them to react jointly to relevant changes in the situation (for example, an unexpected surge in traffic);
(e) To be responsible for creating a situation in which any affected party may employ an appropriate dispute resolution mechanism to identify and seek to correct any underlying problem; and
(f) To be aimed at requiring mutual efforts to ensure the earliest possible correction of the problem and removal of the freeze.[10]

It is worthy of note that the above framework of future regulatory arrangements was intended to function in different structures and relationships, e.g. bilaterally between two States, between a State and a group of States and between two groups of States and multilaterally with a small or large number of States. It was expected that this structure would also respect all rights, existing and newly granted.[11]

---

[10] AT Conf/4-WP/7; 14/4/94, at 3.
[11] See generally, AT Conf/4-WP/16; 23/6/94.

Airlines are faced with the imminent prospect of the future realm of commercial aviation being controlled by a group of air carriers which may serve whole global regions and operated by a network of commercial and trade agreements. Regional carriers will be predominant, easing out niche carriers and small national carriers whose economics would be inadequate to compare their costs with the lower unit costs and joint ventures of a larger carrier. It is arguable that a perceived justification for open skies or unlimited liberalization exists even today in the bilateral air services agreement between two countries, where, *fair and equal opportunity to operate* air services is a *sine qua non* for both national carriers concerned. This has been re-interpreted to mean *fair and equal opportunity to compete* and later still, *fair and equal opportunity to effectively participate* in the international air transportation as agreed.[12] Of course, there has been no universal acceptance of this evolution in interpretation and carriers and States whose nationality such carriers have maintained their own positions tendentiously.

ICAO has suggested the following preferential measures for the consideration and possible use of its member States who are at a competitive disadvantage when faced with the mega trends of commercial aviation and market access:

(a) The asymmetric liberalization of market access in a bilateral air transport relationship to give an air carrier of a developing country: more cities to serve; fifth freedom traffic rights[13] on sectors which are otherwise not normally granted; flexibility to operate unilateral services on a given route for a certain period of time; and the right to serve greater capacity for an agreed period of time;

(b) More flexibility for air carriers of developing countries (than their counterparts in developed countries) in changing capacity between routes in a bilateral agreement situation; code-sharing to markets of interest to them; and changing gauge (aircraft types) without restrictions;

(c) The allowance of trial periods for carriers of developing countries to operate on liberal air service arrangements for an agreed time;

(d) Gradual introduction by developing countries (in order to ensure participation by their carriers) to more liberal market access agreements for longer periods of time than developed countries' air carriers;

(e) Use of liberalized arrangements at a quick pace by developing countries' carriers

(f) Waiver of nationality requirement for ownership of carriers of developing countries on a subjective basis;

(g) Allowance for carriers of developing countries to use more modern aircraft through the use of liberal leasing agreements;

(h) Preferential treatment in regard to slot allocations at airports; and

---

[12] Wassenbergh (1996) at p. 80.

[13] The right to uplift or discharge passengers, mail and cargo in a country other than the grantor State.

(i) More liberal forms for carriers of developing countries in arrangements for ground handling at airports, conversion of currency at their foreign offices and employment of foreign personnel with specialized skills.[14]

These proposed preferential measures were calculated to give air carriers of developing countries a head start which would effectively ensure their continued participation in competition with other carriers for the operation of international air services. Furthermore, improved market access and operational flexibility are two benefits which are considered as direct corollaries to the measures proposed.

The other consideration is the inconsistency between the open skies practice, which while being economically expedient, would undoubtedly phase out smaller carriers who are now offering competition in air transport and a larger spectrum of air transport to the consumer, and safeguarding the interest of developing States. Lower fares, different types of services and varied in-flight service profiles are some of the features of the present system. It is desirable that a higher level of competitiveness prevails in the air transport industry, and to achieve this objective, open competition would play a major role.

8.1.6 Taxation

On the subject of taxation ATConf/6 concluded that the air transport industry has, in recent years, witnessed the proliferation of various types of taxes and levies. This trend, coupled with the lack of transparency, unfair tax schemes, increase in existing taxes to make up for budget deficits and discriminatory practices against air transport vis-à-vis other modes of transport, accordingly are causing serious concern within the industry, and will have a negative impact on the sustainable development of air transport, ultimately affecting national economic development, in particular in developing countries notwithstanding the political and financial difficulties faced by many States and the resulting pressure on regulators, airport operators and air navigation services providers, policy makers and national governments are encouraged to stop the further proliferation of taxes. It was recognized that ICAO has clear policies on taxation and user charges, which remain valid and States should be urged to apply these policies in regulatory practices, in accordance with Assembly Resolution A37-20, Appendices E and F. ICAO should also continue to take the necessary measures to enhance States' awareness of its policies on taxation and user charges and promote application more vigorously.

The Conference agreed that States should give consideration to the following regulatory arrangement to include in the Template Air Services Agreement (TASA) Article on Taxation as an option for use at their discretion in air services agreements:

---

[14] See *Study on Preferential Measures for Developing Countries*, ICAO Doc AT-WP/1789; 22/8/96 at A-7–A-9.

> . . .. Each party shall undertake to reduce to the fullest practicable extent and make plans to eliminate as soon as its economic conditions permit all forms of taxation on the sale or use of international air transport, including such taxes for services which are not required for international civil aviation or which may discriminate against it.

This clause is an option for use by States at their discretion. States may instead choose to use the arrangement in a Memorandum of Understanding (MoU) or a Memorandum of Cooperation (MoC). The Conference recommended that States apply ICAO policies on taxation in regulatory practices, in accordance with Assembly Resolution A37-20, Appendix E. Since ICAO has clear policies on taxation, which remain valid, States should, while avoiding double taxation in the field of air transport, ensure that the policies are followed by relevant authorities in charge of taxation and that they States should avoid double taxation in the field of air transport. States should also, in consultation with all stakeholders, avoid imposing any unjustified or discriminatory taxes on international aviation, and reduce or eliminate any such existing taxes, as they have a negative effect on the competitiveness of the aviation industry and impact States' national economies. ICAO was requested to continue to take the necessary measures to enhance States' awareness of its policies on taxation and promote application more vigorously; and collaborate with relevant industry associations to develop analysis and guidance to States on the impact of taxes and other levies on air transport.

8.1.7 Economics of Airports and Air Navigation Services

ATConf/6 was advised that many changes had occurred in the past two decades in ownership and management in the provision of airports and, to a lesser extent, air navigation services. The changes have generally been described as "privatization." However, these changes can take various forms, and while they generally reflect a move away from government ownership and management, they do not necessarily (indeed rarely) denote outright privatization, particularly with regard to ownership. In addition, changes in ownership and control can also have implications on the governance and performance of airports and air navigation services providers (ANSPs).

As part of its work in this area, ICAO updated and released the ninth edition of *ICAO's Policies on Charges for Airports and Air Navigation Services* (Doc 9082) in April 2012. Based on experience gained worldwide, Doc 9082 reveals that where airports and air navigation services are operated by autonomous entities, their overall financial situation and managerial efficiency have generally improved (paragraphs 4–7 of Section I refer). Therefore, Doc 9082 recommends that where it is economically viable and in the best interest of providers and users, States should consider establishing autonomous entities to operate their airports or air navigation services. Furthermore, it reminds States that when considering the commercialization or privatization of airports or ANSPs, it should be borne in mind that States are ultimately responsible for safety, security and economic oversight of these entities, and States should ensure that ICAO policies are observed and that all relevant obligations, notably as signed in air services agreements, are fulfilled. It is noteworthy that for the economic oversight of providers, specific policies are contained in Doc 9082 related, in particular, to consultation with users, performance management and cost basis for implementation of charges.

A significant development in this respect is that States are encouraged to incorporate the four key charging principles of non-discrimination, cost-relatedness, transparency and consultation with users into national legislation, regulations or policies, as well as into air services agreements. In the context of privatization, these steps are recommended in order to ensure compliance with these key principles by airport operators and ANSPs, regardless of the organizational business model. When considering privatization or private participation in the provision of air navigation services, a cautious approach is required because of cross border and other implications. Detailed guidance material on ownership, control and governance of ANSPs is included in Chapter 2 of the *Manual on Air Navigation Services Economics* (Doc 9161), while more details about the above-mentioned options, their main features and advantages can be found in Chapter 5 of the *Manual on Privatization in the Provision of Airports and Air Navigation Services* (Doc 9980).

Doc 9082 recommends that where it is economically viable and in the best interest of providers and users, States should consider establishing autonomous entities to operate their airports or air navigation services. Furthermore, it reminds States that when considering the commercialization or privatization of airports or ANSPs, it should be borne in mind that States are ultimately responsible for safety, security and economic oversight of these entities, and States should ensure that ICAO policies are observed and that all relevant obligations, notably as signed in air services agreements, are fulfilled. It is noteworthy that for the economic oversight of providers, specific policies are contained in Doc 9082 related, in particular, to consultation with users, performance management and cost basis for implementation of charges.

Doc 9980 updates the circular *Privatization in the Provision of Airports and Air Navigation Services* (Cir 284), which was published in March 2002. Doc 9980 presents information on developments taking place in various parts of the world in ownership and management in the provision of airports and air navigation services, while it provides definitions and analyses of the various available options and possible implications. It also discusses major issues to be examined by States when considering a change in ownership and management. Guidance contained in Doc 9980 takes into account the wide range of circumstances faced by providers of airports and air navigation services and brings to the attention of States other ICAO policy documents related to regulatory measures ensuring the appropriate safeguards in the context of airports and air navigation services management.

The Conference deemed that the current ICAO policies and guidance on charges were relevant, including those on airport and air navigation services funding through user charges. Many States expressed the view that there was no need to develop further guidance as it was considered both premature and redundant to develop a new manual on financing the air transport system. Further, it was expressed that creating a new manual with the intention to complement the existing guidance material would likely be counterproductive and hinder ICAO's efforts to promote the use of key funding principles contained in existing guidance.

It was noted that the future air navigation system will be complex and very costly, and that there would be a need to develop new financing mechanisms. Support was expressed for work to be undertaken on the schemes of economic incentives, "best equipped, best served" and "most capable, best served" concepts, and performance-based approaches from a gate-to-gate perspective. Some States were of the view that ICAO should develop new guidance material to address such issues, while others felt that the relevance of the recently updated guidance material contained in Doc 9082 and in the forthcoming fifth edition of the *Manual on Air Navigation Services Economics* (Doc 9161) should be first assessed.

There was consensus on the need to establish a multi-disciplinary working group composed of ICAO, representatives of States and all other interested parties with a view to further considering these issues. It was suggested that the working group could, initially, seek to determine the parameters and definitions of, for example, service priority policies and ascertain the effectiveness of these policies. Subsequently, the working group could consider how the policies might be applied in practice and, finally, how they could be reflected in existing ICAO policies and other guidance material. The working group could present its findings to the Airport Economics Panel (AEP) and the Air Navigation Services Economics Panel (ANSEP) to determine if and how the existing guidance could be amended to incorporate the findings of the working group. It was finally noted that determining whether such practices are consistent with ICAO's policy on non-discrimination was necessary.

The Conference concluded that ICAO policies and guidance on funding air transport infrastructure through airport and air navigation services charges, through pre-funding of projects and through the allocation of global navigation satellite system (GNSS) costs remain valid and that States should continue to implement ICAO policies and guidance that can be applied to funding air transport infrastructure through airport and air navigation services charges. It also recognized that the relevance of the recently updated ICAO policies and guidance material contained in *ICAO's Policies on Charges for Airports and Air Navigation Services* (Doc 9082) and the *Manual on Air Navigation Services Economics* (Doc 9161) should be assessed in the context of the modernization of the air transport system. Finally, the Conference thought it appropriate to establish a multi-disciplinary working group in cooperation with States, international organizations and the industry in order to consider the economic and operational challenges associated with the air navigation services upgrades in particular, and with financing the air transport system in general.

8.2 Implementation of ICAO's Policies and Guidance

The Conference unanimously recognized ICAO's leadership role in the economic regulation of international air transport and in the development of comprehensive policy guidance. States expressed support for ICAO's activities with respect to updating, enhancing and promoting its policies, guidance and other material related to economic regulation. There was general support for the need to keep ICAO principles, policies and guidance current and responsive to the changing situations and requirements of States. There was also general support for ICAO to continue to promote awareness and explore means by which to enhance the status of its policies for wider use and adherence by States.

Regarding a possible new Annex to the Chicago Convention, there was no consensus on its development. However there was general recognition of the usefulness of such an Annex to enhance the implementation of ICAO policies and guidance. One view expressed that the new Annex may impinge on State sovereignty and be at odds with the need for States to liberalize air transport at their own pace. The different nature between standards and recommended practices (SARPs) adopted in the technical field and ICAO guidance in the air transport field was also mentioned. Other States supported the idea of establishing a new Annex but considered that such an undertaking would require further consideration. Many delegates requested that this issue be considered jointly by ICAO and States with a view to conducting analyses on the scope and content of a new Annex.

The Conference also unanimously supported the need to establish priorities for ICAO's work in the air transport field. Regarding the practicalities of such a prioritization, different views were expressed. Some States considered that the Conference, in its earlier deliberations, had established the following areas as priorities: market access, air carrier and ownership, fair competition and consumer protection. Others suggested that the issues of infrastructure financing and taxation should also be considered as priorities, stating that all ATConf/6 agenda items are of equal importance and interrelated. It was suggested that these priorities should be established on the basis of usefulness, time constraints, complexity of tasks, and available resources. Finally, it was agreed that air transport is a priority for the Organization and that the new ICAO Strategic Objective, namely Economic Development of Air Transport, has equal status in importance as all other ICAO Strategic Objectives.

It was also pointed out that during the Twelfth Air Navigation Conference (ANConf/12) and the High-Level Conference on Aviation Security (HLCAS), no prioritization had been carried out. To assist ICAO in this prioritization, it was suggested that a State letter be disseminated requesting views on the priority issue. Considering the limited time available, the proposal was not considered realistic within the existing time frame leading to the 38th Session of the ICAO Assembly.

The Conference concluded that air transport is a priority, as already recognized by the Council which adopted a new Strategic Objective: Economic Development of Air Transport. The leadership role of ICAO in economic regulation of international air transport, and in the development of comprehensive policy guidance to assist States in the creation of a favourable regulatory environment for the sustainable development of air transport and for the benefit of all stakeholders, is indisputable. ATConf/6 also recognized that ICAO policies and guidance material on the economic aspects of international air transport regulation remain relevant but need to be kept current and responsive to the changing situations and needs of States and aviation stakeholders. It suggested that ICAO should, in cooperation with the industry ensure widespread awareness and improved implementation of its policies as well as use of its guidance material on economic regulation; and ICAO should conduct analyses on the scope and content of a possible Annex to the Chicago Convention on sustainable economic development of air transport, and to work jointly with States on the usefulness and feasibility of such an Annex, including exploring other acceptable solutions.

The Conference concluded that States should be encouraged to recognize the importance and relevance of ICAO policies and guidance and give due regard to them in regulatory practices and that they should be encouraged to exert all efforts to ensure adherence to commitments relating to provisions of Assembly Resolutions in the air transport field. Another recommendation was that States should take into account ICAO principles, policies and guidance in national legislations, rules and regulations, and in air services agreements. The Conference recommended that ICAO perform four major functions in this regard:

(a) Continue to promote its policy guidance on the economic regulation of international air transport, and encourage States to use such guidance in their regulatory practice;
(b) Ensure that policies, guidance and other material related to economic regulation remain relevant, current, and responsive to changing situations and requirements of States;
(c) Continue to consider additional ways and means by which to enhance the status of its policies for the sustainable development of the air transport system, and should assess the value of a possible new Annex to the Chicago Convention on economic development of air transport, or other acceptable solutions; and
(d) Establish priorities according to Recommendations agreed by the Conference.

There are some significant but inconsistent statements here. The first is that "The Conference unanimously recognized ICAO's leadership role in the economic regulation of international air transport". If ICAO has a leadership role in regulating international air transport, why is it not allowed to regulate? Why is the role of the Organization still only, as the statement which follows says: "in the development of comprehensive policy guidance"? How does a leader lead by just producing guidance material? The farthest the Conference went was to acknowledge that States support ICAO's activities with respect to updating, enhancing and promoting its policies, guidance and other material related to economic regulation. The most perplexing statement issued at the end of the Conference was that finally, the Conference agreed that air transport is a priority for the Organization and that the new ICAO Strategic Objective, namely Economic Development of Air Transport, has equal status in importance as all other ICAO Strategic Objectives. If the new Strategic Objective is equally important as the other three Strategic Objectives, why is ICAO restricted to continuing its role, which it has had for 67 years, of just producing guidance material?

8.3   A New Annex?

An encouraging recommendation of the Conference was that ICAO should continue to consider additional ways and means by which to enhance the status of its policies for the sustainable development of the air transport system, and should assess the value of a possible new Annex to the Chicago Convention on economic development of air transport, or other acceptable solutions. The more interesting part of this statement is "or other acceptable solutions". What are these other acceptable solutions? And why could not the Conference have deliberated on them? Could not the organizers have given a few more days for the Conference to discuss these important issues? Why was the 12th Air Navigation Conference, which is held once every 10 years, given 10 working days for discussion, and ATConf/6, which also occurred after 10 years, given only 5 working days?

As for the new Annex on air transport economics, it would be ludicrous to argue that such an Annex would dilute and obviate ICAO policy and guidance material. There are numerous other Annexes, such as Annex 18 on the carriage by air of dangerous goods to name just one, which are accompanied by policy guidance material. There is also Annex 17 on security which is considered an air transport subject at ICAO, which has an accompanying Security Manual, not to mention another air transport Annex—Annex 9 (Facilitation) which is also accompanied by the Facilitation Manual. It is even more ludicrous to suggest that provisions in an Annex on liberalization of air transport would impinge on sovereignty of States. Sovereignty and liberalization are two separate issues. Besides, sovereignty is not what it used to be and connotes State responsibility rather than a right.[15]

It is quite reasonable to think that subjects such as market access and ownership and control are still not ripe enough to go into an Annex as the former would constitute a clash with Article 6 of the Chicago Convention[16] and the latter would

---

[15] For an in depth discussion on this subject see Abeyratne (2013), at Chapter 2. See also by the same author, Abeyratne (2012) at pp. 15–16.

[16] Article 6 stipulates that no scheduled international air service may be operated over or into the territory of a contracting State, except with the special permission or other authorization of that State, and in accordance with the terms of such permission or authorization.

clash with the national legislation of States and standard bilateral practice. However, there are some core principles, particularly in the field of airports and air navigation services charges that could be included in a new Annex. There are also some provisions in *Policy and Guidance Material on the Economic Regulation of International Air Transport*[17] which are candidates for inclusion in an Annex. For example, consideration could be given to the guidance material on the avoidance or resolution of conflicts over the application of competition laws to international air transport[18]; Aircraft Leasing; Air Cargo[19]; and consumer interests,[20] just to name a few.

ICAO's Policies on Charges for Airports and Air Navigation Services[21] contain the recommendations and conclusions of the Council resulting from ICAO's continuing study of charges in relation to the economic situation of airports and air navigation services provided for international civil aviation. There are numerous provisions that could be considered for inclusion in a new Annex. To start with Paragraph 9 is a good example. It provides that States should permit the imposition of charges only for services and functions which are provided for, directly related to, or ultimately beneficial for, civil aviation operations; and that States should refrain from imposing charges which discriminate against international civil aviation in relation to other modes of international transport. Paragraph 12 provides that where economically viable and in the best interest of providers (airports and ANSPs) and users, States should consider establishing autonomous entities to operate their airports or air navigation services, recognizing that in some circumstances a single entity may operate both airports and air navigation services, and that the entity may be in the form of an autonomous civil aviation authority.

One could well argue that these provisions could remain as they are and where they are without being imported into a fresh document. However, there are two fallacies linked to this argument: firstly, policies and guidance material are quite different from Annex provisions which are subject to Articles 37 and 38 of the Chicago Convention which require States to adhere to Standards and Recommended Practices in Annexes and for States to advise ICAO if they are not able to adhere to Standards. There is no such requirements pertaining to policies and guidance materials; secondly, as are provisions in all other Annexes, the provisions of a new Annex would also be subject to being monitored and reviewed, even audited. This would not only give ICAO the much vaunted a "leadership" role but would also ensure that States take the economic regulation of air transport seriously.

Yet another provision that could get into an Annex is in paragraph 13 with regard to private involvement where it says that States, when considering the

---

[17] ICAO *Doc 9587*, Third Edition—2008.
[18] *Id.* Appendix 2 at A2-1 to A2-7.
[19] *Id.* A4-7.
[20] *Id.* A4-11.
[21] *Doc 9082*, Ninth Edition—2012.

commercialization or privatization of airports and ANSPs, bear in mind that the State is ultimately responsible for safety, security, and, in view of the potential abuse of dominant position by airports and ANSPs, economic oversight of their operations. In Paragraph 15 international cooperation through a regional approach in the provision and operation of air navigation services is encouraged in instances where this is beneficial for the providers and users concerned and with a view to facilitating the efficient and cost-effective implementation of the ICAO Global Air Traffic Management (ATM) Operational Concept on the basis of the guidance provided in the *Global Air Navigation Plan*.[22]

Paragraph 16 of Doc 9082 provides that States should review the governance structure of their airports and ANSPs and ensure the use of best practices of good corporate governance, as applicable, with regard to: objectives and responsibilities of the entities; shareholders' rights and their treatment; responsibilities of the board; power and accountability of management; relationship with interested parties; and disclosure of information. In order to promote transparency, efficiency and cost-effectiveness in the provision of an appropriate quality of services and facilities, airports and ANSPs should apply the principles of best practices in all areas of their business. This is yet another potential Annex provision.

As per paragraph 18, States should exercise their economic oversight responsibilities clearly separated from the operation and provision of airports and air navigation services, with roles and powers clearly defined for each function. The main purpose of economic oversight should be to achieve a balance between the efforts of the autonomous or private entities to obtain the optimal effects of commercialization or privatization and those public policy objectives that include, but are not limited to, the following:

(a) Minimize the risk of airports and ANSPs engaging in anti-competitive practices or abusing their dominant position;
(b) Ensure non-discrimination and transparency in the application of charges;
(c) Ascertain that investments in capacity meet current and future demand; and
(d) Protect the interests of passengers and other end users.

 To promote these objectives, consistent with the form of economic oversight adopted, States should ensure that airports and ANSPs consult with users and that appropriate performance management systems are in place.[23] Also for consideration could be paragraph 22 which provides that within their economic oversight responsibilities, States should ensure that providers develop and implement appropriate performance management systems. Paragraph 25 on consultation is also valid for consideration. ICAO's Policies on charges, which appear in Part II of Doc 9082

---

[22] Doc 9750.

[23] Doc 9082, *supra,* note 21, paragraph 19.

could remain separately from the Annex as guidelines or be appended to the Annex as they are.

Another area of consideration, as to whether a new Annex could accommodate taxation principles, is reflected in the entirety of Council Resolution on Taxation of International Air Transport as contained in *ICAO's Policies on Taxation in the Field of International Air Transport*.[24]

References

Abeyratne R (2012) Aeronomics and law. Springer, Heidelberg
Abeyratne R (2013) Convention on international civil, aviation – a commentary. Springer, Heidelberg (Chapter 2)
Wassenbergh H (1996) De-regulation of competition in international air transport. Air Space Law XII

---

[24] Doc 8632, Third Edition—2000 at 3–5.

Chapter 9
Conclusion

Perhaps the most significant lapse of ATConf/6 was its inability to define ICAO's role. Had the Secretariat of ICAO proposed an Annex on air transport, the ensuing discussion may have prompted States to finally define the role of ICAO in the economic field instead of tepidly referring to the word "leadership". This lack of courage and leadership on the part of ICAO paved the way for the Conference to conveniently dismiss the notion of an Annex on the basis of an inept working paper which was badly drafted and presented by a regional organization. The basis of ICAO's innovation and creativity at ATConf/6 was restricted to the fact that the ICAO Secretariat, in its working papers to the Conference, proposed over 50 discrete tasks for ICAO in the nature of projects and activities in the air transport area which let the member States dictate once again what the tasks ICAO should undertake over the next 10 years.

As discussed earlier in this book, it is also regrettable that there was no specific working paper from the Secretariat on ICAO's role in the economic field, given the introduction of a new Strategic Objective in the area. This clearly shows that ICAO had not given adequate consideration as to the manner in which it would implement the Strategic Objective.

Ironically, ICAO has a fine example to follow in regard to its possible approach and role in air transport in the nature of the Organization's Air Navigation Plan (GANP) and resultant Aviation System Block Upgrades, as already mentioned in the book. These ASBUs are not mandatory but provide direction for States in their role in ensuring safety as well as recognizing ICAO's leadership in the air navigation field. ASBUs also assist ICAO in determining the direction in which States are headed and in monitoring States. The ASBU modules are organized into flexible and scalable building blocks that can be implemented depending on the operational needs, while recognizing that implementation of a particular module is not mandatory in all areas or circumstances. The approach adopted is not limiting and recognizes that deployment in addition to the material described in the ASBUs may also take place or be necessary.

From a legal and regulatory perspective, ASBUs are viewed in the context of their main purpose—as modules that are developed and designed to achieve

harmonization and interoperability that result in improvements in the provision of
air navigation services globally. Harmonization in this context is consistency in
procedures and practice. Since not all ASBUs are intended to be implemented
everywhere simultaneously, they have to be prioritized in terms of their implementation to assist in the determination as to the circumstances, places and timeframes
in which they should be implemented. Such prioritization has to be done firstly at
State level and then on a regional basis. Analogically, modules in the air transport
field could be developed that would reach towards harmonization in connectivity.
Such material could well be accommodated in an Annex as Recommended
Practices that would distinctly have more legitimacy and effect than mere guidance
materials.

The following discussion on ASBUs could be applied analogically in the context
of modules or practices in the field of air transport. Given the above, it is clear that
the Council has no coercive power over States in their implementation of ASBUs.
The Council, under the Convention, has only functions (which are in essence
duties) and no powers.[1] On the other hand the Assembly has powers and duties
accorded to it in the Chicago Convention,[2] one of which is to delegate to the
Council the powers and authority necessary or desirable for the discharge of the
duties of the Organization and revoke or modify the delegations of authority at any
time.[3]

Article 54 n) provides that the Council can consider any matter relating to the
Convention which any Contracting State refers to it, giving the Council the capacity
to make its own determination and recommendations pertaining to a matter referred
to it. In this context, the Council may invoke Article 44 which identifies as one of
the aims and objectives of ICAO the developments of techniques and principles
pertaining to air navigation, and discuss States' implementation of ASBUs in the
context of the GANP.

A significant issue in the determination of ICAO's effectiveness as an international organization is the overriding principle of universality and global participation of all its 191 Contracting States in the implementation of ICAO policy. This
principle, which has its genesis in the Chicago Conference of 1944, has flowed on,
gaining express recognition of legal scholars. This is what makes ICAO unique as a
specialized agency of the United Nations and establishes without any doubt that
ICAO is not just a tool of cooperation among States.

The ICAO Symposium on Global Runway Safety,[4] which was held from 24 to
26 May 2011, brought to bear this global leadership role of ICAO, together with the
fact that it is time to define roles and have the legislative courage to give effect to a

---

[1] Although Jacob Schenkman, in his well documented and logically reasoned treatise on ICAO
states that "The Council has been entrusted with duties, powers and functions. . ." he does not give
a single example of such a power. See Schenkman (1955) at p. 158.

[2] Article 49 of the Chicago Convention.

[3] Article 49 h) of the Chicago Convention.

[4] See Abeyratne (2011), pp. 427–440.

safety reporting culture and the legal legitimacy given to ICAO by the Chicago Convention. In this context, there are a few home truths one has to consider with regard to the Council. The first is that the 36 members of the Council are voted in by the Assembly of 191 States. Therefore the Council members, who are not sent in by their States on the States' own accord, are obliged to act in the best interests of their constituents, which are all States comprising ICAO. It would therefore be morally reprehensible for a member of the Council to act in the interests of its own.

Notwithstanding the Council's limited scope on the implementation of ASBUs, the Council could robustly monitor their implementation. Article 55 (c) requires the Council to:

> Conduct research into all aspects of air transport and air navigation which are of international importance, communicate the results of its research to the Contracting States, and facilitate the exchange of information between Contracting States on air transport and air navigation matters.

This could be tied to the objective of ICAO to meet the needs of the people of the world for safe, regular, efficient and economical air transport. In this regard the Council should initiate studies that involve research into safety of air navigation, which is of paramount international importance. This is provided for in the Chicago Convention. Such studies, taking into account global, regional and national implementation trends, could analyse their effects on the improvements on safety regionally and globally. This in turn could result in a compendium of implementation of safety measures for States.

The first question that has to be asked is, is non-implementation of ASBUs by a State which has a need for their implementation, a breach of a legal obligation? In other words, do ICAO member States have legal obligation to implement ASBUs? As already discussed, ASBUs come under the philosophy of Article 37 of the Chicago Convention where States undertake to collaborate in harmonizing regulations with each other, and in the case of ASBUs would do so using the PBN concept. PBN is governed by Assembly Resolution A37-11. Therefore it would be too simplistic to conclude that States could totally ignore ASBUs and consider their implementation unnecessary. ASBUs are tools whereby States would ensure that ICAO performs its functions under Article 44 of the Chicago Convention.[5] To ICAO has been attributed international personality and therefore

---

[5] Article 44 provides: "The aims and objectives of the Organization are to develop the principles and techniques of international air navigation and to foster the planning and development of international air transport so as to: (a) Insure the safe and orderly growth of international civil aviation throughout the world; (b) Encourage the arts of aircraft design and operation for peaceful purposes; (c) Encourage the development of airways, airports, and air navigation facilities for international civil aviation; (d) Meet the needs of the peoples of the world for safe, regular, efficient and economical air transport; (e) Prevent economic waste caused by unreasonable competition; (f) Insure that the rights of contracting States are fully respected and that every contracting State has a fair opportunity to operate international airlines; (g) Avoid discrimination between contracting States; (h) Promote safety of flight in international air navigation; (i) Promote generally the development of all aspects of international civil aeronautics.

it is not merely an aggregate of its member States and it has a right to expect performance of its practices and procedures. The International Court of Justice, in the *Reparations for Injuries Case* held:

> There is an undeniable right of the [sic] Organization to demand that its members shall fulfil the obligations entered into by them in the interest of the good working of the Organization.[6]

This obligation in the context of ASBUs is reflected in ICAO Assembly Resolution A37-15 (Consolidated statement containing ICAO policies and associated practices related specifically to air navigation) Appendix L of which relates to implementation of regional plans. The Appendix recognizes that in accordance with Article 28[7] of the Chicago Convention, Contracting States undertake, insofar as they may find practicable, to provide air navigation facilities and services necessary to facilitate international air navigation. It also acknowledged that the Regional Plans set forth the requirements for facilities and services for international civil aviation; and any serious deficiencies in the implementation of Regional Plans may affect the safety, regularity and efficiency of international air operations and therefore should be eliminated as quickly as practicable.

The Assembly resolved that priority should be given in the implementation programmes of Contracting States to the provision, including continuing operation, of those facilities and services the lack of which would likely have a serious effect on international air operations; the identification and investigation of and action by the Organization on serious deficiencies in the implementation of Regional Plans be carried out in the minimum practicable time; and regional planning and implementation groups shall identify problems and shortcomings in Regional Plans and in the implementation thereof, along with suggested remedial measures.

ASBUs are a compelling and critical link to implementing regional plans implemented through the PIRGs. In this context the Assembly resolved that the Council take into account the requirement to improve still further existing safety levels, should inform fully and promptly each Contracting State of the recommendations for the provision of air navigation facilities and services that are applicable to that State under the Regional Plans. Contracting States are required to prepare and keep up to date suitable plans, including the requirements for personnel, for the orderly implementation of the parts of Regional Plans applicable to them.

---

[6] 1949 ICJ Reports at 184.

[7] Article 28 provides: "Each contracting State undertakes, so far as it may find practicable, to: (a) Provide, in its territory, airports, radio services, meteorological services and other air navigation facilities to facilitate international air navigation, in accordance with the standards and practices recommended or established from time to time, pursuant to this Convention; (b) Adopt and put into operation the appropriate standard systems of communications procedure, codes, markings, signals, lighting and other operational practices and rules which may be recommended or established from time to time, pursuant to this Convention; (c) Collaborate in international measures to secure the publication of aeronautical maps and charts in accordance with standards which may be recommended or established from time to time, pursuant to this Convention.

Furthermore, the Council is requested by the Resolution to arrange for the monitoring of the status of implementation of the Regional Plans and for the issue of periodic progress reports which should include information on serious shortcomings in implementation of the Regional Plans. The users of air navigation facilities and services are required to report any serious problems encountered due to the lack of implementation of air navigation facilities or services required by Regional Plans. The reports should be addressed to the Contracting States responsible for implementation. These States should act on such reports to resolve the problems, but when remedial action is not taken users should inform ICAO, through the medium of an international organization where appropriate. Finally Appendix L requires that the Council arrange for periodic review of serious problems encountered by users due to the lack of implementation of air navigation facilities or services and, when appropriate, for measures to be taken to facilitate elimination of the problems as quickly as practicable.

An earlier Resolution of the ICAO Assembly, adopted at the 22nd Assembly (Montreal, 13 September to 4 October 1977), brings to bear the importance of ASBUs (which were not envisioned at the time). Resolution A22-19 (Assistance and advice in the implementation of regional plans) resolved that ICAO give high priority to fostering and implementing regional plans and called on member States to Contracting States should note the possibility of using operating agencies as a means of fulfilling their international obligations under Article 28[8] of the Convention. It also calls upon States to examine with other States in the region whether the implementation of the particular Regional Plan could be facilitated through bilateral or multilateral agreements; and to hold informal meetings, whether initiated by Contracting States or convened by the Secretary General, which are confined to implementation problems affecting two or more States, should be encouraged where no other effective and timely means are available to resolve the problems.

The ICAO Council is required to render assistance to Contracting States in planning and developing those portions of national implementation programmes related to the provision of facilities and services called for by Regional Plans; and promote implementation of those parts of the Regional Plans with which they are concerned, and, furthermore, the Council should ensure that the Regional Offices are utilized to the fullest extent possible in the carrying out of these tasks.

The Council is also called upon to ensure that all the activities of the Organization that can contribute to the implementation of Regional Plans are carefully

---

[8] Article 28 provides: "Each contracting State undertakes, so far as it may find practicable, to: (a) Provide, in its territory, airports, radio services, meteorological services and other air navigation facilities to facilitate international air navigation, in accordance with the standards and practices recommended or established from time to time, pursuant to this Convention; (b) Adopt and put into operation the appropriate standard systems of communications procedure, codes, markings, signals, lighting and other operational practices and rules which may be recommended or established from time to time, pursuant to this Convention; (c) Collaborate in international measures to secure the publication of aeronautical maps and charts in accordance with standards which may be recommended or established from time to time, pursuant to this Convention.

coordinated, in particular at the regional level. This provision is directly applicable to the implementation by States of ASBUs. The Resolution also provides that when a Contracting State, having explored all methods and means for implementing the Regional Plans with which it is concerned pursuant to Article 28 of the Convention, experiences difficulties which hinder such implementation, it should report accordingly to ICAO and, with respect to those items that might become serious deficiencies if not implemented, it should request assistance from ICAO; and the Council should continue, as a matter of priority, to assist and encourage Contracting States to meet their responsibilities under Article 28 of the Convention, and to investigate the practicability of any other solutions for obtaining implementation of specific facilities and services determined by the Council to represent serious deficiencies in the world air navigation network. What was established and adopted as a Resolution by the Assembly in 1977 has seen the light of day in the form of ASBUs and it is the obligation of States to implement ASBUs when necessary, so that their ultimate obligation of adhering to Standards and Recommended Practices (SARPs) of the technical Annexes to the Chicago Convention is met.

In the ultimate analysis, ASBUs are an integral part of safety management. Safety Management Systems (SMS) are processes which proactively manage the projected increase in aircraft incidents and accidents brought about by the increase in air traffic movements, and ASBUs are calculated to achieve this goal. The upcoming 38th ICAO Assembly Session in late 2013 should, in formulating and adopting an Annex on ASBUs to the Consolidated statement containing ICAO policies and associated practices related specifically to air navigation, ascribe a compelling role to the ASBU, thus avoiding presenting it to the aviation community as a mere option to be applied at the discretion of States. The Resolution should be so contrived and structured that a compelling link is drawn between ASBUs on the one hand and SARPs, Performance Based Navigation (PBN) the GANP, Global Aviation Safety Plan (GASP) and the Global Air Traffic Management Operational Concept on the other.

Analogically, had ATConf/6 taken the initiative to develop specific direction that would have moved toward a Resolution at the 38th Session of the Assembly, it would have achieved the unique result of providing more direction to ICAO in its role in implementing the new Strategic Objective.

It is inevitable that some states may lack the necessary resources to implement ASBUs when the need arises. Therefore some degree of assistance or a mechanism must be in place. At the airport or the ANSP level, ICAO's policies[9] state that costs directly related to the safety oversight function for airport services or for air navigation services may be included in the airport or ANSP cost basis for user charges at the States' discretion and provided that such costs are imposed on the providers of services.[10] It is also stated in ICAO policies that policies and charging principles contained therein can be applied in other circumstances, such as for

---

[9] *ICAO's Policies on Charges for Airports and Air Navigation Services* (Doc 9082).

[10] *Id.* paragraph 2 x) of Section II and paragraph 3 vii) of Section III.

funding of the safety oversight function at the national and the regional level. In all circumstances the four key charging principles of non-discrimination, transparency, cost-relatedness and consultation with users, as well as the requirement that users should only pay for services received, should be observed with respect to funding for safety oversight.

Leaving the example of ICAO's GANP and the ASBUs aside, another compelling factor that cries out for a proactive role for ICAO and some degree of harmonization for a globally liberal regime through the aegis of ICAO is the sheer advantage of "opening up" the market. Cross border interactions, particularly in the context of connectivity of cities along the lines enumerated in this book, would undoubtedly yield more economic benefits when flows of capital and people are opened up. Also, from an aviation perspective, quite apart from the economic benefits and in addition to them, the compelling cultural value of opening up would endorse the purpose of aviation—that of connecting people towards global friendship and understanding.

ICAO is a political institution like any other international Organization. It is a forum, as indeed its Vision and Mission Statements say—an arena which serves for political interaction between member States, their Representatives on the Council as well as the Assembly, to engage in persuasion and the exercise of power. In the process there could be political bargaining leading to commitment among members. According to one commentator, international Organizations enhance the credibility of commitments because they "raise the costs of deception and irresponsibility".[11] On this basis, an Annex, rather than mere guidance material of ICAO would represent tangible commitments by States who would be compelled to demonstrate that they can be counted on to keep to their institutional commitments.

Finally, one is compelled to note, as mentioned earlier in this book, that ATConf/6 recognized ICAO as the only forum for initiating global solutions for the development of a sustainable air transport system. When this statement is tied to the opening speech at ATConf/6 of the President of ICAO's Council, Roberto Kobeh González who said:

> Given aviation's historic role in supporting improved social development and economic prosperity, the potential benefits of growth are enormous over the coming decades. There is a serious chance these benefits won't fully materialize, however, unless we come up with practical and concrete recommendations for adapting the global regulatory framework to the realities of the 21st century.[12]

If ATConf/6 did its part to come up with recommendations, ICAO has now to live up to its role as "the only forum for initiating global solutions for the development of a sustainable air transport system".

---

[11] Keohane (1984), at p. 97.

[12] http://www.icao.int/Newsroom/Pages/once-a-decade-ICAO-air-transport-conference-convenes.aspx.

References

Abeyratne R (2011) Ensuring global runway safety: a look at the future. Air Space Law 36 (6):427–440

Keohane R (1984) After hegemony: cooperation and discord in the world political economy. Princeton University Press, Princeton

Schenkman J (1955) International Civil Aviation Organization. Librairie E. Droz, Geneve

Index

A

Abu Dhabi-based airline, 93
African aviation
  African civil aviation history, 22
  airports and air navigation services, 21
  career and opportunities, 24–25
  NEPAD, 23–24
  SADC, 23
  technical cooperation projects, 19–20
  Yamoussoukro Declaration, 22–23
African Civil Aviation Commission
      (AFCAC), 88
Air cargo
  facilitation, air transport, 82–83
  security (*see* Cargo security)
Airline connections
  airport slots
    airline alliances, 72
    airport capacity, 67
    Article 15, 69–70
    ATC authority, 69
    ATConf/5, 71
    capacity-constrained airports, 65, 68, 70
    dispute settlement mechanism, 72
    EU and USA, 66
    historic precedence, 66
    IATA WSG, 66
    ICAO, 67–68
    Level 3 airports, 65–66
    night curfews, 73–76
    ownership of, 70
    policy approaches, 67
    slot allocation process, 66
  Declaration
    clause 4.4, 58
    clause 6.1, 59

  community of interest, 58
  customers and public service, 58
  international treaty, 59
  safety and security, 58
  definition, 59
  tourism
    aircraft engine emissions, 65
    business travel, 59
    climate change mitigation, 64
    employment, 64
    greenhouse gas emission, 64
    Least Developed Countries, 63
    leisure activity, 63
    leisure travel, 59
    low cost carrier, 60, 65
    market conditions, 65
    transit visa processes and policies, 60
    UNWTO (*see* United Nations World
        Tourism Organization (UNWTO))
Airline Deregulation Act, 165
Airlines
  air cargo services, 136–137
  air navigation services
      (*see* Air navigation services)
  air transport liberalization, 136
  ATConf/5, 137
  Cargo security (*see* Cargo Security)
  charges (*see* Charges, airlines)
  Chicago Convention, 136
  connectivity
    ATConf/6, 106, 107
    global economic recession, 106
    hub and spoke model, 107
    IATA, 103–104
    LACAC, 103
    ownership, 105

R. Abeyratne, *Regulation of Air Transport*, DOI 10.1007/978-3-319-01041-0,
© Springer International Publishing Switzerland 2014

secondary slot trading, 106
slot allocation system, 104–106
World Bank forecast airline, 103
consumer rights
Dhabi-based airline, 93
Emirates, 93, 94
profitable business, 92
QANTAS, 94
Qatar, 93
Singapore, 93
economic aspects, security, 138–139
economic regulation and security
ICAO, 137–138
risk based approach
capacity building, 144–146
ICAO Risk Content Statement,
143–144
insider threats (see Insider threats)
tax
aircraft equipment tax, 108
aviation fuel tax, 108
contribution, 109
definitions, 108
international air transport, 109
legal definition, 109
pecuniary contribution, 108
public burden, 110
Air navigation services
airborne equipment, 152–153
airspace user/airport operator, 150–151
benefit, 151
CANSO, 150–151
Chicago Convention, 153–154
economics of, 194–196
flight capability upgrade decisions, 152
non-capable flight, 152
operational and/or financial measures, 152
performance metrics, 152
service priority policy, 152
Air Navigation Services Economics Panel
(ANSEP), 113
Airport Economics Manual (Doc 9562), 114
Airports Council International (ACI), 86, 107,
116–117
Air traffic control (ATC) slot, 69
Air traffic management, 171
Air Transport Regulation Panel (ATRP), 184,
185, 189
Air Transport Regulatory Panel (ATRP/11),
28–29
Antifragile, 157
Arab Civil Aviation commission (ACAC), 68
Aviation Security Panel (AVSECP), 184

Aviation System Block Upgrades (ASBUs), 5,
168–169
antifragility, 168
Article 55 (c), 169
Article 37 of Chicago Convention, 205
ICAO Assembly, 207
International Civil Aviation Organization,
5, 168–169
PBN concept, 205
Regional Plans, 206
safety management, 208
States' implementation, 204

C
Cargo security
air cargo industry, 139
civil aviation, 139
destroy aircraft, 140
ICAO, 139–140
mass destruction, weapon of, 142
risk assessments, 141
security risk assessment, 141
sustainable aviation security, 141
CEANS. See Conference on the Economics
of Airports and Air Navigation
Services (CEANS)
Charges, airlines
CEANS 2008 (see Conference on the
Economics of Airports and Air
Navigation Services (CEANS))
definition
Chicago Convention, 110
civil aviation, 111
Contracting State, 110–111
Council Statements, 111
ICAO, 112–113
non-aviation purposes, 111
particular industry, 110
property, 109
ICAO's policies (see International Civil
Aviation Organization (ICAO))
marginal cost pricing
ANSConf 2000, 115
economic pricing principles, 115
economic principles, 114
ICAO's pricing strategy, 113, 115
Study Group, 113
single till concept
ACI, 116–117
aeronautical and non-aeronautical
revenues, 115–116
aircraft operators, 116

airports, 119
ANSConf 2000, 117
autonomous business enterprise, 118
economic implications, 117
IATA, 117
ICAO policy, 116, 120
national aircraft, 120
non-aeronautical income, 116
non-profit operation, 119
pooling airport revenues, 116
privatization process, 119–120
Chicago Convention, 7
airlines, 136
air navigation services, 153–154
Article 37 of Chicago Convention, 205
charges, airlines, 110
consumer rights (see Consumer rights)
facilitation, air transport, 78–79
ICAO
Article 28, 206
Article 44, 204, 205
Article 82, 51
Article 83, 51
safeguard measures, 30
low fare business model, 163
Civil Air Navigation Services Organization
(CANSO), 150–151
Civil Aviation Authority, 86
Clayton Act, 91
Code of Federal Regulations (CFR), 88
Competition strategy
advantages, 12
airline ownership and control
air carrier designation and
authorization, 39
air carriers, 42
anti competitive practices, 44
ATConf/5, 40
bilateral air services agreement, 42–43
international cooperation, 44
licence, 43
nationality clause criteria, 39
Regulation 2407/92 of EU, 43
transnational investments, 40
U.S negotiation, 45
ATConf/6, 11
classical theory, 11, 54
definition, 11
foreign investments, 11
franchising and code sharing
agreements, 36
free market competition, 38
fuel prices, 36

globalization, 54
government, role of, 12
home base, 12, 55
ICAO (see International Civil Aviation
Organization (ICAO))
ICC, 38
laws and policies, 13
leisure market calls, 37
liberalized market, 13
market shares, 37
production, factors of, 11
United Arab Emirates, 53
Computer Assisted Passenger Pre-screening
System, 5
Conference on the Economics of Airports
and Air Navigation Services
(ANSConf 2000), 112, 114, 115
Conference on the Economics of Airports and
Air Navigation Services (CEANS)
air navigation services economics and
management, 123–124
airport economics and management,
122–123
airports and air navigation service, 122
consultation with users, 122
ICAO, 121–122
recommendations of
airport charges, 126
airport infrastructure, 127
categorizing services, cost allocation
purposes, 127–128
consultation process, 125–126
cost recovery, security measures, 127
differential charges, 126–127
economic oversight, 124–125
economic performance and service
requirements, 125
governance, ownership and control,
126, 127
ICAO member States, 128
incentives, 128
KPAs, 125
rate of returns, 126
Connectivity. See Airline connections
Consumer protection
AFCAC, 88
ATConf/5, 85
ATConf/6, 86–88, 184–186
Australian Competition and Consumer
Commission, 92
Civil Aviation Authority, 86
CJEU, 90
Draft Convention, 87

Consumer protection (*cont.*)
    European rules, 85
    European Union (EU), 89
    LACAC, 86
    pricing, 91
    principle of equal treatment, 90
    United States competition law, 91
    UNWTO, 86–88
Consumer rights
    Airbus Industrie, 97
    airlines
        Dhabi-based airline, 93
        Emirates, 93, 94
        profitable business, 92
        QANTAS, 94
        Qatar, 93
        Singapore, 93
    air transport industry, 95, 100
    aviation's strategic direction, 94–95
    civil aviation, 101
    FDI, 97
    global aviation community, 94
    hyper-fragmented global industry, 97
    IATA Annual General Meeting, 96
    ICC, 99
    international aviation community, 99
    predictions, transport planning, 98
    in Qatar, 93
    in United Kingdom, 96
Court of Justice of the EU (CJEU), 89, 90
Cross-border trade, 10

D
Dhabi-based airline, 93
Displacement Theory, 142
Dubai Airports, 93

E
Emirates Airline, 93, 94
e-Passport, 81
Essential Service and Tourism Development
    Route (ESTDR), 80
Ethiopian Airlines, 21
European Civil Aviation Commission
    (ECAC), 7
European Guarantee Funds' Association for
    Travel and Tourism (EGFATT), 86
European Union (EU), 7
    airport slots, 66
    ATConf/6, 187

    community carrier principle, 41
    consumer protection, 89
    ICAO, 7
    Regulation 2407/92, 43
    Regulation No. 261/2004, 86
e-Visa, 80

F
Facilitation, air transport
    air cargo operations, 82–83
    Annex 9, 77
    biometric data, 81–82
    Chicago Convention, 78–79
    cost–benefit analysis, 77
    ePassport, 81
    passengers
        ATConf/6, 79
        economic benefits, 79
        ESTDR, 80
        G20 economies, 80
        visa requirements, 79
    Recommended Practice 3.9, 81
    Standards and Recommended Practices on
        Facilitation, 77
Facilitation Panel (FALP), 184
Fifth Worldwide Air Transport Conference
    (ATConf/5), 137
    airline ownership and control, 40
    airport slots, 71
    consumer protection, 85
    low fare business model, 163–164
    night curfews, 74
Foreign direct investment (FDI), 97, 182

G
Game Theory, 167–168
General Agreement on Trade in Services
    (GATS), 60, 177
Gesetz gegen Wettbewerbsbeschra"nkungen
    (GWB), 91
Global Air Navigation Plan (GANP), 5,
    128, 201

H
Harmonization, 6
High Level Conference on Aviation Security
    (HLCAS), 138–140, 197
Hotels, Restaurants & Cafe's in Europe
    (HOTREC), 86

I

IATA. *See* International Air Transport
    Association (IATA)
ICAO. *See* International Civil Aviation
    Organization (ICAO)
Insider threats
    air cargo, 148
    Conference, 146–147
    continuous risk assessments, 147
    ICAO member States, 150
    known shipper and consignor, 148
    security controls, 146, 149
    security culture, 149
    TSA, 148
    USAP, 150
International Air Transport Association
    (IATA), 6–7, 87
    airline ownership and control, 104
    airlines, connectivity, 103–104
    airport slots, 66
    charges, airlines, 117
    IATA Annual General Meeting, 96
    ICAO
        liberalization states, 16
        moral leadership, 17
        multilateralism, 17
        stakeholder, 14–15
        United Nations, 16
    marginal cost pricing, 114
    Slot Conference, 75
International Automobile Federation (FIA), 86
International Chamber of Commerce (ICC),
    38, 99
International Civil Aviation Organization
    (ICAO), 27, 46–47, 203
    African aviation (*see* African aviation)
    air cargo security, 5–6
    airport charges, policy
        Declaration, 129
        Doc 9082, 130
        Legal Status, 134–136
        regulatory provisions, 131–133
    air transport connectivity, 6–7
    Arab Civil Aviation commission, 68
    Article 44, 27
    Article 57, 1
    Article 28 of Chicago Convention, 206
    Article 44 of Chicago Convention, 204, 205
    ASBUs (*see* Aviation System Block
        Upgrades (ASBUs))
    Assembly Resolution A37-15, 206
    ATConf/6 (*see* Sixth Air Transport
        Conference (ATConf/6))

ATRP/11, 28–29
AVSEC Mechanism, 4
Computer Assisted Passenger
    Pre-screening System, 5
ECAC, 7
economic measures, 170
economic regulation, 2
epiphenomena
    ATConf/6, 159
    economics field, 159
    ESTD scheme, 160
    global competition, 161
    national prosperity, 161
EU, 7
European Transport and
    Telecommunications Council, 4–5
facilitation (*see* Facilitation, air transport)
free trade, 9
GANP, 5
goal of, 1
governance system (*see* African aviation)
IATA, 6–7
    liberalization states, 16
    moral leadership, 17
    multilateralism, 17
    stakeholder, 14–15
    United Nations, 16
ICAO Air Services Negotiation Conference
    (ICAN), 177
*ICAO Comprehensive Aviation Security
    Strategy* (ICASS), 145
ICAO Contracting State, 111
ICAO/ECAC Model Clauses, 104
*ICAO's Assistance and Capacity Building
    Strategy for Aviation Security*, 145
*ICAO's Policies on Charges for Airports
    and Air Navigation Services*, 111,
    125, 196, 200
*ICAO's Policies on Taxation in the Field
    of International Air Transport*, 202
9/11 issues, 4
*Manual on Air Navigation Services
    Economics*, 113
market place, 9, 10
Mission Statement, 3
PBN, 205
policy and guidance material
    Article 82 of Chicago Convention, 51
    Article 83 of Chicago Convention, 51
    Doc 8632, 48
    Doc 9626, 48
    leadership role, 49–50
    multilateral approach, 50

SARPs, 47
State implementation, 49
Vienna Convention, 51
political institution, 209
Regional Plans, 207
Resolution 33/1, 4
safeguard measures
Article 43 of Charter, 32
ATConf/6, 30
Chicago Convention, 30
"flag carrier" concept, 31
foreign airline services, 30
foresight-awareness culture, 33
fuel prices, 34
global economy, 35
global financial system, 34
jointventure operations, 30
market access, 32
relief flights and humanitarian law, 31
research and innovation strategies, 34
technology and market trends, 33
United Nations, 31–32
safety measures, 205
Strategic Objectives, 3
12th Air Navigation Conference, 5
6th Air Transport Conference, 5, 6
UN, role of, 3–4
WHO/NGO relations, 17
world trade, 10
International Hotel & Restaurant Association
(IH&RA), 86

K
Kenya Airways, 21
Key performance areas (KPAs), 125

L
Latin American Civil aviation Commission
(LACAC), 86, 103
Least Developed Countries (LDCs), 63
Low fare business model
aircraft types, 163
airline–airport relationship, 163
Airline Deregulation Act, 165
ATConf/5, 163–164
capital management, 166
carriage, by air, 162
cheap transportation, 162–163
Chicago Convention, 163
hub and spoke system, 161
ICAO Air Transport Conference, 164
legacy carriers, 165, 167

market conditions, 162
market deregulation, 166
network management, 162
partnerships and global alliances, 162
simple services and process, 163

M
Manual on Air Navigation Services Economics
(Doc 9161), 196
Manual on Air Navigation Services Economics
(Doc 9161/3), 114
Memorandum of Cooperation (MoC), 194
Memorandum of Understanding (MoU), 194
Multilateral Agreement on the Liberalization
of International Air Transportation
(MALIAT), 175, 179–180

N
New Partnership for Africa's Development
(NEPAD), 23–24
Night curfews
aircraft noise, effects, 73
ATConf/5, 74
capacity-constrained airport, 74–75
ICAO Balanced Approach, 74
WSG, 75–76

O
One Stop Security, 147
Oxford Economics, 35

P
Performance Based Navigation (PBN), 205
Policy and Guidance Material on the
Economic Regulation of
International Air Transport, 200
Privatization in the Provision of Airports
and Air Navigation Services
(Cir 284), 195

S
Security to the Convention on International
Civil Aviation, 138
Sherman Act, 91
Sixth Air Transport Conference
(ATConf/6), 203
air carrier ownership and control
airlines, 181
effective control, 183

facilitate liberalization, 180
FDI, 182–183
international agreement, 182
national airlines, 180
policy guidance, 181–182
sclerotic, 182
waiver, 181
airline connectivity, 106, 107
air navigation services, 194–196
airports, economics of, 194–196
air transport economics, 199
alliances, 173
ANSPs, 201
competition strategy, 11
consumer protection, 86–88, 184–186
facilitation, air transport, 77, 79
fair competition
    airline alliances, 186
    Common Aviation Area, 187
    core principles, 188
    European Union, 187
    global air transport market, 187
    international air services, 188
    liberalization or deregulation, 186
ICAO, 27
    epiphenomena, 159
    policies and guidance, 197–199
    safeguard measures, 30
market access
    air cargo liberalization, 178
    air cargo operations, 175
    bilateral, regional and plurilateral
        arrangements, 176
    GATS, 177
    ICAN, 177
    international agreement, 175
    international air transport liberalization,
        176–177
    liberalization, air transport, 175
    MALIAT, 175, 179–180
    pursue liberalization, 176, 177
national and regional regulatory
    frameworks, 174
policy guidance material, 199
public policy objectives, 201
safeguards, liberalization process
    airlines, 192
    aviation community, 190
    commercial aviation and market access,
        192–193
    effective participation, 190

international air transport system, 189
market share argument, 191
policy guidance, 190
safety net, 191
slot allocation process, 66
stakeholders, 173
States and relevant international
    organizations, 173
TASAs, 174
taxation, 193–194
taxation principles, 202
UNWTO, 60
SkyTeam, 33
South African Airways, 21
Southern African Development Community
    (SADC) countries, 23
Standards and Recommended Practices
    (SARPs), 4, 47, 143, 154, 197, 208
Star Alliance, 33

T
Template Air Services Agreement (TASAs),
    174, 193
The Black Swan, 157
Tourism, 5, 21
    aircraft engine emissions, 65
    business travel, 59
    climate change mitigation, 64
    employment, 64
    greenhouse gas emission, 64
    Least Developed Countries, 63
    leisure activity, 63
    leisure travel, 59
    low cost carrier, 60, 65
    market conditions, 65
    transit visa processes and policies, 60
    UNWTO (see United Nations World
        Tourism Organization (UNWTO))
    World Travel and Tourism Council, 79
Trade Practices Act, 92
Transportation Security Administration
    (TSA), 148
Twelfth Air Navigation Conference
    (ANConf/12), 5, 197

U
United Federation of Travel Agents'
    Associations (UFTAA), 86
United Nations Educational, Scientific and
    Cultural Organization (UNESCO), 17

United Nations World Tourism Organization
        (UNWTO), 59, 87, 184, 185
    air transport market and liberalization, 60
    ATConf/6, 60
    economic growth, 60
    environmental sustainability, 60
    GATS, 60
    green house gas emission, 64
    ICAO, 88
    internationl tourist arrivals, 63
    role of, 64
    UNWTO World Tourism Barometer, 63
    visa process, 80
    WTO, 60
Universal Security Audit Programme
        (USAP), 150

V
Vienna Convention, 51, 111

W
World Association for Professional Tourism
        Training (AMFORT), 87
World Customs Organization (WCO), 83
World Meteorological Organization
        (WMO), 17
World Tourism Organization (UNWTO), 86,
        87, 184
World Trade Organization (WTO), 10
World Travel Agents Association Alliance
        (WTAAA), 86
World Travel and Tourism Council
        (WTTC), 86
Worldwide Slot Guidelines (WSG), 66, 75

Y
Yamoussoukro Declaration, 22–23
Yaounde Treaty, 22

Lightning Source UK Ltd.
Milton Keynes UK
UKOW04n0708240614

233903UK00001B/44/P